PENGUIN BOOKS

THE CASE OF THE MISSING NEUTRINOS

Dr John Gribbin trained as an astrophysicist at the University of Cambridge before becoming a full-time science writer. He has worked for the science journal *Nature* and the magazine *New Scientist* and has contributed articles on science topics to *The Times*, the *Guardian* and the *Independent*, he has also made several acclaimed science series for BBC Radio 4. John Gribbin has received awards for his writing in both Britain and the United States and is currently a visiting Fellow in astronomy at the University of Sussex. In 1999 he was elected Fellow of the Royal Society of Literature. His many books include *In Search of Schrödinger's Cat*, *Stephen Hawking: A Life in Science* (with Michael White) and *In Search of SUSY*. John Gribbin is also the author of several science fiction works, including *Innervisions*.

He is married with two sons and lives in East Sussex.

ALSO BY JOHN GRIBBIN IN PENGUIN

In Search of the Big Bang
In Search of the Edge of Time
In Search of the Double Helix
In Search of SUSY
In the Beginning
The Stuff of the Universe *(with Martin Rees)*
The Matter Myth *(with Paul Davies)*
Stephen Hawking: A Life in Science *(with Michael White)*
Richard Feynman: A Life in Science *(with Mary Gribbin)*
The Little Book of Science

THE CASE OF THE MISSING NEUTRINOS

and Other Curious Phenomena of the Universe

JOHN GRIBBIN

PENGUIN BOOKS

PENGUIN BOOKS

Published by the Penguin Group
Penguin Books Ltd, 27 Wrights Lane, London W8 5TZ, England
Penguin Putnam Inc., 375 Hudson Street, New York, New York 10014, USA
Penguin Books Australia Ltd, Ringwood, Victoria, Australia
Penguin Books Canada Ltd, 10 Alcorn Avenue, Toronto, Ontario, Canada M4V 3B2
Penguin Books (NZ) Ltd, Private Bag 102902, NSMC, Auckland, New Zealand

Penguin Books Ltd, Registered Offices: Harmondsworth, Middlesex, England

First published in the USA by Fromm International 1998
First published in Great Britain with a revised introduction and bibliography
in Penguin Books 2000
3 5 7 9 10 8 6 4 2

Printed in England by Clays Ltd, St Ives plc

contents

Introduction

▼

THESE ESSAYS WERE FIRST published in the *Griffith Observer*, over a span of about twenty years, between the mid-1970s and the mid-1990s. I started writing them as light relief from my day job at the science journal *Nature*, where some of the more off-the-wall aspects of astronomy that have always intrigued me were considered not quite the thing for discussion in those sober pages. Some of the essays were written for the annual competition organized by the Griffith Observatory in Los Angeles (and sponsored, at that time, by Hughes Aircraft), which was an intriguing challenge because the essays were judged anonymously – the judges did not know the names of the authors of the essays they were judging. By trial and error, I was able to learn what kind of story and what kind of presentation appealed to an American West Coast audience; as a result, I started using both the competition essays and other contributions to the magazine as test beds for my ideas, trying out bits and pieces that would later be used (sometimes in greatly modified form, sometimes not) in my books. So some of the items here may produce a feeling of *déjà vu* if you have read some of my other books; but there are others which have never previously had a life outside the pages of the *Griffith Observer*.

I have strung the stories together to present a picture of the Universe working outwards from the Earth, where we live, and the Solar System, not in the order they were written. The rough edges have been smoothed a little, but the discerning reader will notice that there are gaps where recent developments might have been referred to if they were being written today. Rather than trying to plug these gaps and risk making the book a patchwork of styles, I have chosen to mention them here, which also gives me a chance to demonstrate the prescience of some of the essays. Since Chapter One was written, for example, the impact crater associated with the death of the dinosaurs has been discovered. Chapter Five deserves a special mention because it was written in 1977, and although I have updated it, it describes a model of solar behaviour that is now out of fashion; but I feel that it helps to highlight how astronomers were groping for a full understanding of solar behaviour less than thirty years ago. The history of science is more messy than some accounts would have you believe!

Chapter Ten is also a little long in the tooth – it dates from the late 1970s. Most of the story still stands up, though; the big new element in the study of galaxy formation in the 1990s is that today cosmologists are absolutely sure that the visible bright matter in the Universe is embedded in a sea of dark matter which holds galaxies in its gravitational grip, and that there is even more of this dark matter than was suggested in the late 1970s. The key difference this makes is that it is easier to make galaxies than people thought in 1980.

Chapter Thirteen is not to be taken entirely seriously today – although even here some of the concepts incorporated in the mathematical description of white holes are very similar to the concepts used to describe the birth of the Universe in the latest theories, called inflation. But Chapter Fifteen, which looks even weirder, deals with what is still very respectable science, and was written in collaboration with Paul Davies, as a spinoff from our book *The Matter Myth*. The rest is all my own work, and the result is very much a personal account of twenty years of watching the Universe, which hits some of the highlights that made headlines in that period, but also highlights some of the stories that were curiously neglected by most reporters. I hope you enjoy it, and I am grateful to the Griffith Observatory for permission to re-use the material in this way.

John Gribbin
July 1999

one

Base Eight Arithmetic, Meteorites and Us

▼

WHY DOES OUR ARITHMETICAL system use the base ten? Obviously, it is because we have two hands each, and each of these hands has five digits on it. There is nothing sacred about base ten arithmetic, however, and if, with a little imagination, we envisage an intelligent life form with four hands each having three fingers, then logically we might expect that life form to count in twelves. On the face of it, assuming that intelligent beings start to count by using convenient parts of their manipulating limbs as markers, there are endless possibilities on which other life forms might base their arithmetic. But how many of these are actually practical possibilities? To start with, would it really make sense to have four arms, each with a three-fingered hand? In evolutionary terms, probably not—at least as far as intelligent life is concerned. Bilateral symmetry, with pairs of limbs on either side of the body, is clearly a successful evolutionary invention. Legs on each side help you to stand up; an odd number of legs couldn't be fitted conveniently into this pattern, although—as animals such as kangaroos have found—a tail can make a useful additional prop.

The number of legs on each side is not so easy to decide, if we are trying to design successful life forms. Many crawling creatures have legs by the dozen (even if the millipede does not actually have a thousand limbs), while very successful species are around on Earth with six or eight limbs. Even so, the ones that use a specialized set of limbs for manipulating things always set aside a *pair* for the purpose. The crab has two large pincers at the front, the kangaroo has two arms, and even a mouse uses its front pair of limbs for holding food. This makes sense. With bilateral symmetry and eyes at one end of the body, obviously one pair of limbs will be most useful for grasping and moving things.

Leaving aside wild speculations about life in the clouds of Jupiter, or on the surface of a neutron star, we might begin, on this basis, to decide just how far our own shape is determined by the conditions under which our ancestors evolved. Are we, in fact, typical of the kind of intelligent life to be found on Earth-like planets? What are the chances that, if ever we do make contact with intelligent beings which have evolved under similar conditions, they too will be upright, bipedal animals with two arms, each ending in a five-fingered hand, and a head mounted on top of the body carrying a pair of eyes, a nose and a mouth?

To start with, intelligent life—the kind that builds civilizations and spaceships—must not be too successful in being adapted to its own natural environment and must have been, during its evolution, under considerable pressure from predators. The elephant is intelligent, by some standards, but so powerful that it is insulated from the dangers of attack by enemies and has never had to use its intelligence to fight off enemies. The whale and dolphin, potentially as intelligent as human beings, also have few enemies and are superbly adapted to their watery environment. The price they have been obliged to pay is streamlining and a total absence of limbs and hands that can be used to manipulate objects. A whale may sing, but could neither construct nor play the cetacean equivalent of a saxophone or piano.

The point is important. Intelligent, tool-using life will emerge on a planet like ours on the land, not in the sea. It won't be very big or fierce, since big, fierce animals have no need to invent tools or weapons, or to sharpen their cunning by hiding from predators. And it will have a pair of limbs at one end, conveniently near the eyes and ending in digits (fingers) that can be used to grasp and manipulate small objects.

The picture already begins to look very much like a primate—a rat-like or squirrel-like creature good at hiding and scurrying away from danger. Sharp eyes and good hearing will detect danger coming, and its size would not be so big that hiding, or running away, becomes difficult. What about more legs for running away with? The centaur, half horse and half man, looks at first like a good bet. But there is a snag. The bigger the body, the harder it is to hide and the more food it needs to survive. A centaur is heavily committed to running, rather than hiding, as a defence, and in evolutionary terms that means that the pressures of natural selection will operate to produce more horse-like centaurs, with the human-like limbs withering away into ever more useless appendages. No, apart from the kangaroo's tail, it is hard to see how we could improve on the basic design of two legs for running, two arms with hands for carrying things, and a head mounted on top carrying two eyes to provide a stereoscopic, three-dimensional view of the world. Three-dimensional vision is essential for judging distances, whether it be the distance to a charging lion or to a morsel of food waiting to be picked up. A good high mounting for the eyes is essential for a prey animal, which needs early warning of impending danger. The necessity of air to breathe and food to eat complete the outline design, requiring a mouth and a nose of some description, although maybe a few variations on those themes are possible.

At a quick glance, even trying to avoid any cultural bias from our everyday experience, it looks very much as if the bipedal design is the right one for intelligent life on Earth. The only real room for variation is in the number of fingers on each hand. Five is certainly a useful number, as we all know. But it does seem to be on the high side of usefulness. It is difficult to see how an extra finger on each hand would be very beneficial, while by contrast many people who have suffered accidents leading to amputations manage very well with only two or three fingers on a hand. The key, in those cases, is that they still have a thumb with which to oppose the remaining fingers, making it possible to grip and manipulate objects dexterously.

So far, all this is speculation. As yet, we have no information about life on other planets with which to test the idea that intelligent Earth-type life is bipedal and, broadly speaking, human-like. What we need is one test case. If we landed one robot probe on one other Earth-like planet and found the dominant life form to be an intelligent biped with four or five fingers on each hand, the argument that this is the inevitable product of evolutionary selection on such a planet would be overwhelming. The chance of such similarities arising by coincidence is so small as to be entirely negligible.

Unfortunately, the chance of landing a robot probe on another Earth-like planet in the immediate future is equally small. But wait—this isn't the end of the story. We do have information about one planet under conditions that were undeniably Earth-like but on which there was no human life. I refer, of course, to the Earth itself, during the era of the dinosaurs. If the arguments I have sketched out above hold any water at all, then the pressures of natural selection, operating during the era of the dinosaurs just as they have in the subsequent 65 million years of Earth history, should have been acting to produce an intelligent, upright biped.

Now, during the 150 million years or so that dinosaurs dominated the Earth, the evolutionary pressures were, in many ways, less than they have been since. In particular, the climate was more stable than it has been in the past few million years, and because of the geographical arrangement of the continents (which changes over millions of years, due to continental drift) there were no great ice ages to weed out species and put a premium on intelligence and adaptability. The recent cycle of ice ages, according to most evolutionary theorists, played a key part in forcing human beings to adapt to changing conditions, putting a premium on intelligence and flexibility and making us what we are today. That is why we have evolved so rapidly.

But even with less pressure on the dinosaurs from the environment, surely intelligence would still have been an advantage. And, surely, over 150 million years even relatively gradual evolutionary changes would have had a chance to get to work.

Indeed they would. Although most people think of dinosaurs as great lumbering brutes with tiny brains, the term 'dinosaur' applies to a variety of creatures as wide as the variety covered by the term 'mammal' today. There were big, stupid dinosaurs, but there were also small, agile dinosaurs. There were meat-eaters—the dinosaur equivalent of lions and tigers, and there were grass-eaters—the dinosaur equivalent of deer and sheep. And the dinosaurs didn't even die out without a trace, whatever the impression most popular accounts provide. Dinosaur descendants are alive and well on Earth today, not just in the form of obviously reptilian creatures like crocodiles and alligators, but also in the form of birds: products of a highly successful dinosaur line that took to the air (as well as developing warm blood, a trick emulated by some other dinosaur lines). Out of all that variety, were there no dinosaur candidates for the bipedal, upright niche that, according to my argument, marks a vital step on the road to intelligence? If the fossil record showed no sign of a dinosaur even remotely human in appearance, we would have to admit that the idea falls down, but if there were dinosaurs that could be described, in the broadest terms, as on the path to human like appearance, then it would at least make the argument look a little more plausible.

In fact, there were several dinosaur types which followed, broadly speaking, the kangaroo's approach to bipedalism rather than the human approach, keeping a large tail which could be used as a stabilizer, weapon, or seat. That's no problem: a biped with a tail is still a biped. *Tyrannosaurus* and *Iguanodon* carried this design to extremes, reaching 5 metres in height One a flesh eater (*Tyrannosaurus*), the other herbivorous, neither of them could be said to be intelligent. *Scleromochlus*, a bipedal reptile about 1 metre long which lived about 200 million years ago, is a more likely candidate for the pre-intelligent niche, but it had a small brain and never seems to have made the grade. But there is a star candidate who fulfils, as far as we can tell from the fossil remains, *all* our requirements. If you landed on a distant planet and were greeted by a creature like *Saurornithoides*, you would have to admit that the argument that Earth-like planets produce human-like intelligent species held water.

Saurornithoides was a smallish dinosaur, weighing about 50 kilograms which lived at the end of the age of the dinosaurs, some 65 million years ago. It had the largest brain, in proportion to body mass, of any dinosaur, with a brain to body weight ratio not far different from that of the modern baboon. And it was clearly an active, bipedal creature, with a long tail behind and four-fingered hands at the end of each arm, the fingers perhaps being arranged as two true 'fingers' with an opposable 'thumb' on either side.

This is a pretty impressive set of credentials. Starting from this basis 65 million years ago, if *Saurornithoides* had followed the same path, in response to similar evolutionary pressures, that the equivalent pre-humans were to follow 60 million years later, then it might well have been possible for a *Saurornithoides* civilization to arise, with eight-fingered, kangaroo-like bipeds developing spaceflight by about 60 million years ago. If so, and if the species had survived whatever unimaginable processes may have faced them over the 60 million years they could have had beyond the present stage of human civilization, the Solar System today might well be the playground of a bipedal society, but one to which base eight arithmetic seemed the obvious choice. Carl Sagan speculated briefly along these lines in his entertaining book *The Dragons of Eden*. But why did the dinosaurs fail to make the breakthrough to intelligence? What stopped *Saurornithoides* from exterminating the mammals and going on to develop their own civilization?

The best answer seems to be that a large meteorite struck the Earth just when these particular dinosaurs were taking the first steps on the road to intelligence, and as a result all the large animals living on the Earth's surface died. This is the explanation for the catastrophe which brought an end to the age of the dinosaurs that is currently in vogue, and it rests upon some very good evidence.

The fact that there was a catastrophe which wiped out all large animals is clear from the geological record. Almost overnight, in geological terms (which means in the space of no more than 100,000 years), half of all the land species on Earth, including all animals bigger than about 40 kilograms in body mass, became extinct. After the disaster, the world was a different place. The surviving small animals, in particular, were now free to move into the ecological niches previously occupied by the large dinosaurs. Most of the small animals moving into those niches were mammals—the small mammals were already well established on Earth during the age of the dinosaurs—and over 65 million years they have evolved into elephants, tigers, gazelles, and so on, replacing the dinosaur equivalents that are now just a fossil memory. If the disaster, whatever it was, had wiped out all animals bigger than 60 kilograms, then *Saurornithoides* would have been well placed to achieve world dominance. As it was, it just missed the boat, and in the fullness of time the little rat-like mammals which had probably been among its prey produced a new intelligent species—but one very much, in terms of superficial appearance, in the *Saurornithoides* mould. Of course, we have no tail, and we do have five fingers on each hand. But we look as much like *Saurornithoides* as we do the tree shrews from which we are descended. On Earth-like planets, it seems that the way to fit the niche for intelligent life is indeed to be bipedal with two arms, two hands and a head mounted on top in the lookout position.

But it also helps to avoid large meteorite impacts. The chances of winning this particular cosmic lottery are not very good, at least in our Solar System. The battered faces of the Moon, Mercury, and some of the moons of Jupiter and Saturn bear mute witness to the frequency of meteorite impacts during the history of the Solar System. Even after the effects of erosion by wind and water, the surface of the Earth shows that such impacts are still hardly rare, on any geological timescale. Barringer Crater, in Arizona, is the classic example. It is 1,200 metres across and 180 metres deep, and was produced by a meteorite impact which can be dated, using standard geological techniques, to only 25,000 years ago. Vastly greater features, such as the West Clearwater Lake in Quebec (21 km across) and the

Vredefort Ring in South Africa (56 km across) show the characteristic circular shape of a meteorite impact and are almost certainly craters produced hundreds of millions of years ago. Clearly, such impacts must have dramatic environmental effects. It is thirty years since Joe Enever presented the now classic calculation of the worst kind of meteoritic disaster, published in an article in *Analog* magazine.

Enever started with simple calculations of the energy involved in producing the Vredefort Ring, using one of the simplest equations in physics—a body of mass m moving at velocity v has a kinetic energy of mv^2, and if that body is brought to a halt by colliding with the Earth, all that energy is liberated as heat. A fairly ordinary meteorite might be moving at 50 km per second when it hits the Earth, and there are bits of such cosmic rubble around in the Solar System with masses of thousands of tonnes. The kinetic energy released by an impact with such a body would yield the equivalent of more than 100,000 megatonnes of TNT, bigger than any nuclear device yet tested.

Even this, however, is not enough to explain the Vredefort Ring, which required an impact yielding 10 million megatonnes equivalent, coming from a collision with an object as big as the minor planet Hermes—32 thousand million tonnes of rock.

If such an object had struck our planet 65 million years ago, it could well explain the demise of the large dinosaurs. Dust blasted high into the stratosphere by the explosive impact would have spread around the Earth like a shroud. The heat itself was, perhaps, the primary agent of death.

The snag with the hypothesis is that there is no crater comparable to the Vredefort Ring in size but only 65 million years old to be found on Earth. But, as Enever pointed out, most of the Earth's surface is covered by water. Suppose the giant meteorite fell in the sea? It might seem that an oceanic strike would be less spectacular than one on land, since it would be 'damped' by the water. But the opposite is the case! Quite apart from incidentals such as the tidal waves that would be produced, the almost unimaginable amount of energy released by the impact would not only vaporize the water of the sea at the point of impact but would punch a hole scores of kilometres wide right through the thin crust of the ocean floor, exposing the hot magma beneath. Sea water pouring into the pit would eventually cool the molten rock and return conditions to normal—but not before 16,000 cubic kilometres of water, by Enever's calculations, had been evaporated in the process.

In this version of the scenario, the Earth would be shrouded by shiny white clouds, reflecting away the Sun's heat, and the water vapour would be precipitated as snow. Perhaps plants would die and animals would starve, with the biggest animals, that need the most food, suffering the worst.

Everything fits. But the idea remained a speculation until 1979, when a team from the University of California, Berkeley, came up with evidence that geological strata 65 million years old are enriched by traces of heavy elements, iridium in particular. The original discoveries were made in strata from Italy; since then, fresh evidence has come in from as far afield as Denmark, New Zealand and the central North Pacific. All the evidence suggests that some global event 65 million years ago—just at the time of the death of the dinosaurs—spread a layer of dust enriched with exotic heavy metals around the world.

The best candidate for such an event is a giant meteorite impact. The Earth's crust is deficient in heavy metals because any that were present when the Earth formed have settled into the dense, molten core. But asteroids, the cosmic rubble left over from the formation of the Solar System, presumably contain a higher proportion of elements such as iridium, since they have no cores into which heavy elements can settle. The amounts found in the key strata are still only traces by any normal standard, but they are equivalent to enrichment of the natural level of iridium in the crust by between ten and a thousand times. Clearly, something happened 65 million years ago, and it would be a remarkable coincidence indeed if that something were not related to the disasters that brought an end to the era of the dinosaurs. The palaeontologists, traditionally a cautious crew, have so far only acknowledged that a giant meteorite impact may have *contributed* to the demise of the dinosaurs, perhaps being the 'last straw' that came after several million years of deteriorating climatic conditions. Whatever, there seems no doubt at all that an event like the one which produced the Vredefort Ring, if it happened tomorrow, would certainly spell the end of *our* civilization, if not of the entire human species (among others).

If it did, though, I'd be willing to make a small hypothetical wager that in 50 or 100 million years' time there would be a species of intelligent bipedal animal doing very nicely on planet Earth. They might not be mammals or reptiles; they might or might not have tails. Maybe they would count in base eight or base ten, but I'd be surprised if they counted in either base six or base twelve. They would be about 2 metres tall, with eyes in heads at the top of their bodies. And they would be speculating about the disaster that brought an end to the age of the mammals, wondering whether the upright bipeds had ever achieved true intelligence, and no doubt joking about the likelihood that those strange creatures with five-fingered hands may have used a bizarre decimal counting system. *Plus ça change, plus c'est la même chose.*

two

The Air We Breathe

▼

ONE OF THE GREAT mysteries in the Universe is the puzzle of life. Is life—even intelligent life—something that arises almost inevitably on planets that are the 'right' distance from their parent stars and contain the 'right' chemical ingredients? Or is it some kind of cosmic fluke, which has occurred only occasionally . . . perhaps only once? We may never know the answers to these questions, for we cannot be absolutely sure of the range of conditions under which life can evolve. All we know for certain is that life has a very firm grip on one particular planet: ours. And here on Earth, the success of life seems to be intimately connected with an abundance of liquid water. Our very term for a region devoid of water, 'desert', is synonymous with a region devoid of life.

So, although we cannot be certain, we make an educated guess that life, more or less 'as we know it', will be common in the Universe if wet planets are common in the Universe. All the other factors—temperature, chemical composition, and so on—are taken care of by one requirement: the presence of large amounts of liquid water. Leaving aside entirely the question of life as we *don't* know it (intelligent jellyfish floating in the atmosphere of Jupiter *à la* Carl Sagan), the question to answer is whether *we* are a rare product of rare astronomical conditions.

As well as being wet, the Earth is a small, rocky planet with an oxygen-rich atmosphere. All these features fit together and relate to our position in the Solar System. The nearness of the Earth to the Sun makes the Earth rocky; the *exact* distance of the Earth's orbit from the Sun has determined the nature of the atmosphere and oceans which cover our rocky planet.

Our Solar System contains two kinds of planet and some bits and pieces of cosmic junk (I include Pluto, the outermost 'planet', in the cosmic-junk category because it is very likely an escaped moon, or some other oddity, and did not form as a planet in its own right). Orbiting relatively close to the Sun there are four rocky planets: Mercury, Venus, Earth and Mars, sometimes called the terrestrial planets because they are all (like the Earth) solid, rocky planets surrounded by only thin layers of atmosphere. Mercury, closest to the Sun, has hardly any atmosphere at all. Farther from the Sun, past the main collection of junk, the asteroid belt, there are four giant planets: Jupiter, Saturn, Uranus and Neptune, which are quite different from Earth-like planets.

For a start, the giant planets are, of course, much bigger. Jupiter, the biggest, has two and a half times as much matter as all the other planets in the Solar System put together and a volume equal to 1,319 Earths. Even Uranus and Neptune, the smallest of the giants, have volumes 60 to 70 times that of our home planet. Most of this great size is made of a sort of liquefied gas. Where the terrestrial planets are solid with a fringe of atmosphere, the giants are fluid with perhaps small rocky cores. And the giant planets are rich in hydrogen, as well as compounds such as methane amd ammonia.

One underlying reason explains these differences—distance from the Sun. When planets formed out of the gas and dusty material that contracted to become the Sun and Solar System, they experienced conflicting forces. Gravity pulled aggregates of matter together, but the increasing radiation from the young star at the centre of the nebula blew them apart, with the lightest, most volatile materials boiling away into space. The heat was, of course, greatest nearest the Sun, where light elements stood literally a snowball's chance in hell of sticking to the planets forming there.

Farther out, where things were cooler, light elements condensed out into gaseous compounds, and gravity held them together to produce giant planets. So, although hydrogen is the most common element in the Universe (and in the Sun and pre-solar nebula), almost all of it was blown out of the inner Solar System. The terrestrial planets are made up almost entirely of leftovers, the tiny percentage of such material present in the original nebula. The Earth itself is composed of elements which were present only as a fraction of 1 per cent of the original nebula, and only a tiny fraction of the Earth's original hydrogen remains. Most of that hydrogen, however, is now found combined with oxygen to produce the wet seas of our planet.

Clearly, the giant planets are nothing like the Earth, so it is no use looking there for seas which might be the breeding-grounds of life as we know it. The four terrestrial planets are different from one another, although they are more like each other than they are like any of the giant planets. And once again the Sun's heat is probably the main factor in these differences. Mercury, the innermost planet, is rather like the core of the Earth stripped of even the elements that might have formed a thick rocky crust. It is very dense, small, and very rich in metals compared with the other terrestrial planets. In addition, it is so hot that it has scarcely a trace of atmosphere. It could never have oceans of running water and can be ruled out as a home for life. The remaining three candidates, though, are much more promising. Venus and Earth, although at different distances from the Sun, are almost the same size as each other and have very similar compositions. Mars, farther from the Sun than the Earth, is a lighter planet (following the general rule that lighter material was dispersed outwards across the Solar System as it formed), but has a respectable atmosphere.

The nature of the rocks beneath our feet is a product of what the heat of the Sun was when the Earth was forming and of the distance of the Earth from the Sun. Much closer, the planet would have little in the way of rocky crust at all; a little farther out, and it might have been simply a rocky core buried deep beneath the thick atmosphere of a gas giant. The nature of the air that we breathe depends even more critically on our distance from the Sun.

Venus, so nearly the Earth's twin in many ways, has a thick blanket of atmosphere, rich in carbon dioxide, which traps solar heat through the 'greenhouse effect' and raises the temperature at the surface to nearly 500°C. Mars has a thin atmosphere, but it is too cold there for water to exist as a liquid or for rain to fall. The Earth alone has an atmospheric blanket which is just right to keep the surface of the planet hotter than the freezing-point of water and cooler than its boiling-point. The result is a wet planet, where water continuously evaporates from the oceans and is recycled as rain—the conditions ideal for life as we know it. How did this come about?

Whatever leftover scraps of light gases were still associated with the terrestrial planets as they formed, they must have been blown away during the erratic activity of the young Sun before it settled down to provide the steady glow we know today. This stabilization took place about 5,000 million years ago. The present atmospheres of the inner planets have been produced from gases that seeped out from their interiors, 'outgassing' from the rocks, including volcanic activity and the vaporization produced when large meteorites hit the surface. It used to be thought that the first atmosphere produced was rich in gases like methane and ammonia, similar to the atmospheres of the giant planets. This idea was tied in with the search for the origin of life, for laboratory experiments had shown that by mixing gases like methane and ammonia with water in a sealed tube, and passing electric sparks or ultraviolet radiation through the mixture, molecules regarded as the precursors of life could be formed. The early Earth was bathed in ultraviolet radiation from the Sun, and the early atmosphere must have provided plenty of sparks in the form of lightning. So the guess was made that methane and ammonia were also present to set life on its way.

More recently, however, other experiments have shown that the prebiotic molecules can be built up in test-tube 'atmospheres' rich in carbon dioxide—and Fred Hoyle and Chandra Wickramasinghe argue that precursors to life are even present in interstellar gas clouds and the material of comets! So there is no longer any need to invoke the presence of a primordial atmosphere rich in methane and ammonia. Instead, atmospheric scientists today argue, logically enough, that the kind of atmosphere produced by the original outgassing must have been rich in the gases that escape from the interior of the Earth, through volcanoes, today. This present-day outgassing produces mainly carbon dioxide and water, and is our best evidence that the early atmosphere of the Earth was mainly carbon dioxide and water. This conclusion is very strongly supported by the discovery that the thick atmosphere of Venus and the thin atmosphere of Mars are both rich in carbon dioxide. Those planets, however, seem to have lost the life-giving water, while we have lost the carbon dioxide. Why? And how?

Everything fits neatly with the orbital distances of the three planets from the Sun. Just about the only simple thing physics can tell us about the conditions at the surface of a rocky planet, at a known distance from the Sun, is the surface temperature. And that, it turns out, is all we need to know! For Venus, the stable temperature at which heat coming in from the Sun is balanced by heat being radiated into space is 87°C (when there is no atmosphere). So, as soon as gas escaped from the rocks and began to build up an atmosphere, it stayed as gas—not just carbon dioxide, but water as well would have stayed in the vapour state. Both water vapour and carbon dioxide trap infrared heat, which is what is radiated by hot rocks. Both gases are transparent to the kind of heat coming in from the Sun, most of which is much shorter in wavelength. So both gases let in the Sun's heat while they stop the heat from the surface from escaping. This is the so-called 'greenhouse effect'. As a result, the initial surface temperature of 87°C rapidly rose as the atmosphere developed. The temperature increased beyond the boiling-point of water and moved toward the conditions we see today. Now a new balance between incoming and outgoing radiation has been struck.

On Mars, things were very different. The surface temperature stabilized at about *minus* 30°C before outgassing began. Water could not even melt, let alone evaporate. Although the thin carbon dioxide atmosphere does produce a greenhouse effect, this is not enough to melt the frozen water today. It is just possible that sometime in the past the atmosphere was thick enough to do the job, and that water did flow on Mars, carving out the canyons and lineated systems which look so much like dried-up river beds. Slight changes in the tilt of the planet towards the Sun could melt the polar caps and perhaps create enough new atmosphere to melt and evaporate sub-surface ice, and so provide a temporarily thicker atmosphere. But as far as we can tell, judging from the number of meteorite craters that scar the 'rivers' of Mars, it has been at least 500 million years since water, or whatever liquid it was, flowed on the surface of our neighbouring planet. Mars is almost exactly on the borderline between being able to maintain life as we know it and not, and it may even be that our descendants will be able to thicken the atmosphere artificially (perhaps by melting the Martian polar caps) enough to make it habitable. Meanwhile, though, Earth remains the odd planet out, the wet one between the extremes of heat and cold.

Here on Earth, the initial surface temperature before outgassing was about 25°C, high enough for liquid water to flow but not so high that enormous quantities of water vapour got into the atmosphere to produce a runaway greenhouse effect. Quite the reverse—the warm waters dissolved carbon dioxide out of the atmosphere, checking the greenhouse effect, so that the new equilibrium the planet reached after an atmosphere formed was actually cooler than it had been without, thanks to the white clouds now reflecting away a good part of the incoming solar heat. The surface temperature of the earth, *with* an atmosphere, settled down at an average of around 15°C, where it has remained ever since, thanks partly to a kind of natural thermostat.

Suppose the Sun warmed up a little, as it might during its lifetime. Instead of the Earth getting hotter and perhaps developing a runaway greenhouse effect, the slight increase in temperature might produce more evaporation from the oceans and more white clouds to reflect the Sun's heat. Or imagine a slight cooling; with less evaporation and fewer clouds, a greater fraction of the Sun's heat—diminished though it would be—would get to the ground and make the cooling less severe than it otherwise would have been. These are just simple hypothetical examples. What matters is that the atmosphere and oceans of our planet have developed in close harmony with the heat arriving from the Sun, which depends, in turn, on the temperature of the Sun itself and our exact distance from it. When life came on the scene, it added a new factor but did not disrupt this balance.

Even after the first atmosphere formed, sterilizing ultraviolet radiation from the Sun was able to reach the surface of the Earth and keep it free from life. But once life developed in the safety of the oceans, it began to play its own part in moulding the environment. The first life forms found oxygen poisonous, a dangerous waste product of their biology, but by a couple of billion years ago oxygen was beginning to build up in the atmosphere. There, chemical reactions stimulated by the Sun's radiation (photochemical reactions) led to the production of ozone, a triatomic form of oxygen, high in the atmosphere, where to this day it acts as a filter which blocks out the sterilizing ultraviolet radiation. Under this protective blanket, life could move out of the sea and onto the land, while an abundance of atmospheric oxygen allowed new life forms to invent the trick of respiration, using the oxygen as an energy source.

As life spread across the land, huge amounts of carbon dioxide were processed by plants into carbon (eventually to be stored up in coal deposits) and oxygen (which fuelled animal life, including our ancestors). Today, the balance is being tipped once again, as our industrial society converts substantial quantities of the carbon back into carbon dioxide through burning fossil fuel. Even worse, at the same time we are rapidly destroying the great tropical forests that are the remaining 'lungs' of our world. The plants in these jungles still use the old-fashioned technique of photosynthesis to break up carbon dioxide into its constituent parts. Perhaps this will eventually lead to a runaway greenhouse effect after all; or perhaps the natural thermostat will take care of the situation. After all, there was no runaway greenhouse 2.5 billion years ago, before life really came to grips with the problem of disposing of the bulk of the carbon dioxide!

But this is a problem for the future. How does this understanding of the origin of the air that we breathe help us to assess the likelihood of life existing elsewhere in the Universe? A surprising number of people seem to take the view that the story I have just outlined means life may be very rare. They argue that there is no evidence that life exists on either of our near neighbours, Venus and Mars, and that we can understand this very well, as I have explained, simply in terms of how far away each of those planets is from the Sun and the temperature of the Sun itself. The Earth seems to be roughly in the middle of the band of orbits around the Sun in which a rocky planet will have the right surface temperature to form a stable atmosphere with lots of running water to make oceans. Venus is definitely too close to the Sun for life as we know it, and Mars is just too far away. So, runs the argument, we are very lucky. With a slightly different orbit, Earth would either freeze, like Mars, or fry, like Venus. The whole argument, however, can be turned on its head. If a star like the Sun has a family of rocky, terrestrial planets, then it is very hard to see how at least one of them could fail to be in an orbit in the temperature band where wet planets exist! Look at the situation from a different viewpoint. Suppose our Sun were just a little bit hotter; then our planet Earth would still be in the viable zone for wet, life-bearing planets, although perhaps a little warmer than it is now. Venus would be just as undesirable as before. But now Mars too would be a wet planet with a reasonably thick atmosphere! And if the Sun were just a little cooler, then Venus and Earth could indeed be sister planets, even though Mars would be a good deal colder. From this point of view, it seems that we are unlucky in this Solar System, having a Sun which is just at the right temperature for one of its terrestrial planets to be wet but not quite right for there to be two habitable planets in the system.

This is a very important shift of viewpoint in terms of the prospects of finding life elsewhere in the Universe. We know, from the evidence that most of the ordinary and common stars rotate slowly, that stars like our Sun always seem to be born together with a family of planets. We know, too, that the same processes which led to two 'families' of planets in our Solar System will work in other systems. It is reasonable to expect at least some of the other stars like our Sun to have a few 'terrestrial' planets in orbits covering about the same range of distances from the central star as are occupied by the orbits of Mars, Earth, Venus and Mercury. And it seems, from the evidence of the near-miss of Mars and the presence of life on Earth, that the habitable band around a star like the Sun—the band in which rocky planets collect oceans of liquid water—is likely to encompass at least one, and possibly two, of its inner planets. The pessimists, who say that if the Earth were a little closer to the Sun we would fry, while if it were a little farther away we would freeze, are excessively gloomy. Almost anywhere from just this side of Venus to just this side of Mars, a planet like Earth will end up wet and habitable! The air that we breathe and the oceans that surround us are not freaks of nature but the inevitable accompaniments to a rocky planet at the corresponding distance from its parent star. There, the equilibrium surface temperature was at least a few degrees above the freezing-point of water before the atmosphere formed. But it was still cool enough for most of the water to remain liquid rather than turn to vapour. The orbital range is small enough to make the chance of any particular rocky planet being wet fairly small. But if the spread of planets in our Solar System is typical, the habitable band is wide enough to make the chances of finding at least one wet planet in each such system fairly good.

Stars like our Sun seem to be ideal breeding grounds for life. The Sun is a long-lived, stable star, with a fairly broad 'life zone', in which a wet planet can exist comfortably for the thousands of millions of years it takes life to evolve. Smaller, cooler stars live even longer, but their life zones are correspondingly narrower. The chances are smaller, therefore, that we shall find a wet planet in just the right orbit around one. Bigger, hotter stars can have very broad life zones, but they go through their own life cycles more quickly than cooler stars. There is insufficient time for the processes which, on Earth, have taken so long to produce us and all the life around us. And this is, perhaps, the happiest realization of all. The conditions which we know are necessary for life as we know it are seen to be common corollaries to stars like the Sun. This suggests that we are not in a special place in the Universe. Wherever stars like our Sun exist, there ought to be planets like the Earth (more or less), with the key feature of free, running water on their surfaces.

So it may indeed be possible, not just that life exists elsewhere in the Universe, but that it thrives on watery planets, with blue skies and white clouds, rivers and trees, and surface temperatures in the range that we are used to here on Earth. This, rather than the science fiction of methane-breathing monsters of intelligent crystalline life forms, inhabitating worlds with purple or green skies and lakes of liquid ammonia, seems to be the simplest interpretation of the evidence. If we are not alone in the Universe, the chances are that the air that we breathe is very similar to the air that 'they' breathe.

three

Waiting for the Next Ice Age

▼

IN 1976 A COMBINATION of several pieces of evidence fell into place. They confirmed climatologists' understanding of the basic cause of the rhythmic ebb and flow of the glaciers that have characterized the more recent series of ice ages here on Earth. Probably, the same processes also affected the changing climate during the previous ice epochs, hundreds of millions of years ago. But the greatest significance of the breakthrough in understanding ice ages is that it provided a firm prediction of how the natural course of events will develop—if undisturbed by our activities—over the next few thousand years. We can now say unequivocally that the warmest period of the present 'interglacial' is over, and that from here on we can expect a cooling-off. Within about ten thousand years the world will be in the grip of another full ice age.

Depending on your point of view, that is either worrying or reassuring. The prophets of doom, comparing the prospect of another ice age with the very long history of the Earth, point out that by those standards the next ice age is 'just around the corner'. On the other hand, optimists such as myself argue that the evidence shows that we do not have to worry about the ice for a couple of thousand years yet. If we survive that long, we should be able to do something about preventing the spread of ice. The astronomical theory of ice ages that makes these predictions possible is generally called the 'Milankovitch model', after Milutin Milankovitch, of Yugoslavia, who spelled out details of the idea more than half a century ago. The basis of the idea is much older, though, and among the others who supported it even before Milankovitch was Alfred Wegener, better known to us today as one of the originators of the concept of continental drift. Like that concept, the Milankovitch model of ice ages has survived for decades before at last being proved basically correct.

Standard textbooks have long acknowledged the existence of the Milankovitch model and spelled out the details of its basic mechanism. Three separate, cyclic changes in the Earth's movements through space combine to produce the overall changes in the solar radiation falling on the Earth, and are the key to the theory. The longest of these is a cycle of between 90,000 and 100,000 years, during which the shape of the Earth's orbit around the Sun stretches from almost circular to something more elliptical and back again. When the orbit is nearly circular, there is a more even spread of solar heating over the year, taking the Earth as a whole. When the orbit is elliptical, we are closer to the Sun at some times than at others. This can increase the contrast between seasons, even though the total heat received by the whole Earth over an entire year may stay the same.

The second effect is a cycle some 40,000 years long, during which the tilt of the spinning Earth changes, the Earth 'nodding' up and down relative to the imaginary line joining the centre of the Earth and the centre of the Sun. This change is known technically as a change in the *obliquity of the ecliptic*, and it directly changes the contrast between seasons. When the tilt is more pronounced there are strong seasonal changes, and when the earth is nearly 'upright' there is less difference between summer and winter.

Finally, the gravitational pull of the Sun and Moon on the bulging equatorial regions of our planet produces a wobble like that of a spinning top—but with a period of 26,000 years. This is the *precession of the equinoxes*.

These effects combine to produce changes in the amount of heat arriving at different latitudes at different times of the year, but they do *not* change the total amount of heat received from the Sun by the whole planet over a whole year. It is very easy to see, in general terms, how this kind of change in seasonal heat could encourage ice to spread, given the present positions of the continents. Cool summers in the northern hemisphere might allow the snow which falls in winter on the land surrounding the polar sea to persist through the summers. Once some snow and ice fields become established in this way, we can imagine that, by reflecting away a good part of the weak summer heat, they will encourage the rapid spread of glaciation through a feedback process.

On the other hand, the conditions we need to produce a spread of ice in the southern hemisphere are just the opposite. What is important there is to have very severe cold winters, in order to freeze more ice from the sea—snowfall alone is no use, since this has little land to fall on, and so a cool summer won't help the ice to spread as much as a severe winter will. What we need for global ice age conditions, then, are cool northern summers plus cold southern winters, and, of course, the two go hand in hand. Since northern summers occur at the same time as southern winters, the astronomical effects that are needed to produce ice ages in both hemispheres also go hand in hand, as long as we have an arrangement of continents roughly like that of the present day. This, of course, raises some interesting long-term questions. Can ice ages only occur when we have this kind of geography?

Any kind of glaciation seems possible only when there is land at high latitudes. Now, it looks as if the combination of a land-locked northern polar sea and a southern polar continent provides the uniquely necessary conditions for the astronomical changes of the Milankovitch model to cause the advance and retreat of the ice on a grand scale. But for years—decades—the unanswered questions have remained. Has the fluctuating pattern of past ice ages actually followed the changes the model predicts? Just which seasons of the year are the critical ones, the ones in which relatively small changes in solar heating trigger the spread of ice and snow cover? And are the changes in solar heat (insolation) really enough to explain the spread of ice or to melt the glaciers at the end of an ice age? With many climatologists eagerly investigating the problem, it was only a matter of time before the answers came, but it was something of a coincidence that all three questions were answered in the same year, 1976.

Part of the reason for this 'coincidence' was the improvement in computer modelling techniques that took place in the early 1970s. One example of the power of the new computer models came when Professor Johannes Weertman, of Northwestern University in Illinois, used the technique in an analysis of how ice sheets grow under the influence of small changes in heat radiation. He found just the kind of changes needed to explain ice ages by the Milankovitch model, confirming that this could be the true cause of ice ages. But as he said, the evidence did not on its own settle the issue once and for all—the results demonstrated 'that the Milankovitch radiation variations are potent enough to be the prime cause of ice ages, although [they] by no means prove that the Milankovitch ice age theory is correct.' Various other attempts have also been made to produce computer models of how ice sheets vary, and they generally produce the same kind of results, showing that the Milankovitch variations could cause ice ages, but not proving that they do.

Another aspect of the modelling approach is to use our present understanding of the way the reflectivity of the Earth changes, and of the changing amount of heat stored up in the oceans, combined with the present geography, to work out a 'global heat budget' and find out which past epochs should have been so deficient in heat—or overdrawn on the budget!—that ice must spread. Dr David Adam of Menlo Park, California, is one of the modellers who tried this approach and he came up with one curious, but perhaps important, result. According to Adam's calculations, northern-hemisphere ice sheets could form without producing a dramatic change in the climate of the southern hemisphere because the land-based ice sheets would have little effect on overall ocean circulation. But changes in the southern ice must affect the whole globe by changing the whole pattern of ocean currents. Adam suggests that perhaps two mechanisms act to cause ice ages—the Milankovitch effect with its more or less regular, periodic changes, together with the less predictable situation of Antarctic ice surges. These surges might themselves be triggered by changes in snowfall and seasonal temperatures produced by the Earth's wobbles through space. But we should remember that not all ice ages need be caused by the Milankovitch mechanism. That prediction, of a new ice age within about 10,000 years, assumes not only that we won't interfere with nature, but also that there will be no surge of the Antarctic ice in the intervening time. Perhaps the prophets of doom would be better occupied watching the southern ice cap than studying the Earth's movement through space.

One of the most succinct summaries of just where this kind of modelling leaves the Milankovitch theory was provided by Drs Max Suarez and Isaac Held, of Princeton University, in a short paper published in *Nature* in 1976. Their model is relatively sophisticated and can be used to predict the temperatures of two layers of the atmosphere (representing the upper and lower halves of the troposphere, where most of the weather takes place), the surface temperatures over land and ocean, the depth of snow over land and the thickness of sea ice, all calculated for different latitudes and different seasons as required. By changing the incoming solar radiation in line with the Milankovitch model, the Princeton team was able to 'predict' how the climate of the past 150,000 years should have varied. This study has particular value because seasonal changes are specifically picked out: for example, changes in the warmth of northern-hemisphere summers are particularly important in affecting the overall climate, just as we suspected. Overall, the 'predictions' do not match the actual record of the past 150,000 years precisely, but the agreement is close enough for Suarez and Held to conclude 'that a substantial portion of climatic variability on these timescales can be understood as the equilibrium response to perturbations in the orbital parameters'—in other words, a good part of the climatic changes are caused by the Milankovitch mechanism. Calculations made by Dr George Kukla of the Lamont-Doherty Geological Observatory, New York State, pinned down the critical season for the insolation changes even more precisely. The critical time is the northern-hemisphere late summer/early autumn, with the interiors of North America and Eurasia seen as the main geographical locations for the Milankovitch influence on snow and ice cover to be felt. But we are still left with the two key questions to answer. Does the record of past ice ages show really convincing evidence of the Milankovitch cycles? And is there really enough energy involved in these fluctuations to do the trick?

The breakthrough in determining just how temperatures and ice sheets have fluctuated over the past few hundred thousand years, published in 1976, was certainly a long time in the making. It is necessary to have long-range, accurate information if you want to test for the reality of a cyclic change as much as 90,000 to 100,000 years long, just as we need several hundred years of evidence to test for climatic effects with periods of a few tens of years. Until recently, long-range information on climate was very sketchy, incomplete and, we now know, downright inaccurate.

In the old picture of ice ages, geologists reckoned that there had been four or five periods of greatly increased northern-hemisphere ice cover in what, to a geologist, could be regarded as the 'recent' past. But new evidence built up since about 1960 shows that in fact there have so far been something like a score of these glacials, separated from each other only by rather short warm periods, or interglacials, during the history of the current ice epoch.

It is hardly surprising, then, that Milankovitch himself, and other supporters of his theories, could not make the model fit the 'evidence' in the 1930s, 1940s or 1950s. The Milankovitch model 'predicted' many more ice ages, much more closely spaced through the recent history of the Earth. The geologists had no evidence for them all. When that was the best evidence available, the Milankovitch model had to be left out in the cold. But as new techniques have provided ever better evidence of how the Earth's climate has changed over the past few hundred thousand years, the picture built up from the evidence has changed repeatedly. Every change has brought the evidence more nearly in line with the Milankovitch model, until now there remains no doubt at all that the periodic ebb and flow of the great glaciers do indeed include cyclic changes in line with the rhythms of the model.

This is not the place to describe at length the revolution in geophysics that included a complete revision of the time-scales attributed to ice ages, but the key technique which led ultimately to the breakthrough in establishing the reality of the Milankovitch cycles is worth spelling out in a little detail.

The technique depends upon two factors: a better supply of raw material from which past climates can be reconstructed, and better methods for carrying out the reconstruction. The first part, the raw material, is now provided by research ships that cruise the oceans of the world, retrieving cores of mud and other debris from different parts of the sea bed. As the years go by, sediments are deposited on the sea bed, building up this ooze in layers which correspond to the passing years, decades and centuries. Old sediments, deposited perhaps hundreds of thousands of years ago, lie deep in the mud, while new sediments, deposited only yesterday, are at the top. And in between lie the sediments from all the years in between. These sediments cannot be interpreted like tree rings—there are no obvious thick and thin layers to interpret in climatic terms. But there is evidence of climatic change to be revealed by modern analytical techniques.

In particular, the techniques used with great success in analysing ice cores can be applied to these ocean sediments. That depends upon measuring the proportion of the heavy oxygen isotope, oxygen-18 present, because this proportion is affected by temperature. For the ice cores, the technique is relatively straightforward since, after all, every molecule of water ice contains its oxygen, which in principle at least can be analysed. But where is the oxygen of the past locked up in ocean sediments for us to study? The answer: in the shells of small marine creatures, the fossil remains of the animals known as foraminifera. These shells are basically made of chalk which contains oxygen atoms locked up in compounds with other elements—oxygen atoms plucked from the air at the time the creatures were alive, and preserved where we can now measure the proportion of heavy oxygen present and use the measurements to work out a record of temperature changes covering hundreds of thousands of years.

There is also a bonus here for the workers trying to pin down this climatic timescale. One of the other principal components of the chalky remains of the foraminifera is carbon, and that carbon too is representative of conditions on Earth at the time the animal was alive. The radiocarbon (carbon-14) dating technique can be applied to these shells, at least for the more recent part of the cores (carbon dating is less effective for longer timescales, but other radioactive tracers can be used to lengthen the 'calendar').

As long ago as 1955, Dr Cesare Emiliani, working in Chicago, established from the oxygen-18 fluctuations in fossil remains from cores drilled in the Caribbean that there had been many more abrupt changes in climate in the past few hundred thousand years than could be fitted into the then-conventional picture of ice ages. Hardly surprisingly, geologists did not welcome this evidence, and it was not until the late 1960s that, with so many of their cherished beliefs overturned by the new understanding of plate tectonics and continental drift, a revolution in the understanding of past climates proved possible. Improved techniques on land also showed many more ice ages than the handful that had hitherto been accepted, and the clinching evidence came when improved dating techniques showed that the evidence of ice ages from the land record did indeed match up in time with the evidence from the oceans.

By the 1970s, the time was right for a new investigation of the Milankovitch model, and techniques had advanced to provide even more evidence about the pattern of past ice ages. As well as the direct measurements of oxygen-18 in the sediments, the number and nature of the fossil shells themselves can tell us about the climate at the time their inhabitants were alive. Just as we do not find polar bears living in tropical regions, or humming-birds in the Arctic, so the different species of foraminifera have their own preferred climates. One family of these tiny animals may prefer colder water, while another prefers slightly warmer water—and by counting the numbers of each animal in different sediment layers it is possible to get another handle on the changing climate. And another new technique allows the experts to work out from the fossil foraminiferal remains how much ice there was over the globe at different times in the past.

As well as the direct effect of temperature on the amount of oxygen-18 getting into the air and being absorbed by living creatures, the amount of ice cover also plays a part. The molecules of water that contain oxygen-18 instead of the lighter oxygen-16 do not move so speedily as the lighter molecules, and as a result it is easier for them to stick onto a growing ice sheet and freeze. So the ice sheets end up with a high proportion of oxygen-18, leaving relatively little to be absorbed by sea creatures (or anything else). At first sight, this looks like a problem. After all, at face value a high proportion of oxygen-18 in the fossils should indicate cooler conditions. But at the same time, if the ice sheet effect has been at work, then we might expect less oxygen-18 in the fossils from ice ages.

The answer is stunningly simple. First, the changing oxygen ratios in creatures that are known to have lived in the deep ocean are measured. Because these creatures are hardly affected at all by temperature changes at the surface, the varying proportion of oxygen-18 revealed can be used to calculate how the ice cover of the globe has changed. Then, with this effect allowed for, oxygen ratios from different species, surface dwellers that are strongly influenced by global temperature changes, can be interpreted as a direct measure of changing temperature—a global thermometer.

With thousands of sea-floor cores available, and two decades of accumulated expertise in interpreting them, Drs Jim Hays, John Imbrie and Nicholas Shackleton put together the most definitive study of the cycles of ice advance and retreat over recent millennia yet available. The results, published in the journal *Science* late in 1976, were based on detailed study of two selected cores, one of which provided a good continuous record of temperature and ice-cover changes going back almost half a million years—sufficient time to test for the presence of even the 90,000—100,000-year cycle.

The various different climatic tests were applied to samples taken from these cores at intervals of 10 centimetres throughout each core's length which combine to give a 15-metre span covering the past 450,000 years—just 150 sample 'dates' in all, providing a spacing of about 3,000 years between each date, which is sufficient to pick up any cycle as short as 20,000 years. So the three Milankovitch cycles were nicely bracketed by the evidence, and all that remained was to test for the existence of the cycles, using the statistical analysis techniques which pick out the 'spectrum' of periods present in any 'signal' that contains many interacting cycles, and some random effects, or 'noise'. Different tests on the different climatic indicators produce slight differences in the detailed numbers that come out of this analysis, but Dr Hays and his colleagues found close agreement between the results, confirming the reality of three significant cycles:

> There can be no doubt that a spectral peak centered near a 100,000-year cycle is a major feature of the climatic record . . . dominant cycles . . . range from 42,000 to 43,000 years [and] three peaks . . . correspond to cycles 24,000-years long.

Of these effects, the shorter-period influences are more dramatic, while the roughly 100,000-year-long cycle of 'stretch' in the Earth's orbit acts to modulate the influence of the shorter cycles, the coldest periods of ice ages coinciding with times when the Earth's orbit is more nearly circular. This way there is no countereffect from the cycle to offset the deepest dips in the other cycles. The roughly 26,000-year-long cycle seems to have had a strong effect in recent millennia, producing changes in climate almost immediately as the astronomical influence changes. And the statistics also reveal a less important 19,000-year cycle, which is explained by the latest astronomical calculations as a secondary effect of the precession-of-the-equinoxes cycle.

In some ways, the 42,000-year period is the most intriguing. Although there is a clear impact of the cycle on climate over a period corresponding to ten of these 'rolls' of the Earth, and including four major ice ages, the climatic cycles seem to follow about 8,000 years after the astronomical cycles. This delayed response can be explained in terms of the delays built into the climatic system of the Earth—the effect, perhaps, of the build-up and subsequent surge of the Antarctic ice sheet, of other, more subtle adjustments of the air, sea and ice to changing outside influences. Hays, Imbrie and Shackleton have no doubts, however, about the conclusions to be drawn from their study:

> It is concluded that changes in the Earth's orbital geometry are the fundamental cause of Quaternary ice

> ages. A model of future climate based on the observed orbital—climate relationships . . . predicts that the long-term trend over the next several thousand years is towards extensive northern-hemisphere glaciation.

Computers and statistical analysis are all very well, but physicists like to be able to explain such patterns of behaviour in physical terms. In this case, just how do the small changes in seasonal heat allow ice to build up so dramatically, or melt ice so quickly? This last piece of the puzzle was slotted into place by Dr B. J. Mason, Director-General of the United Kingdom Met Office, with calculations of classic simplicity. They are so convincing that, having set out with the avowed intention of proving the Milankovitch model wrong, Mason ended up convincing himself that it is correct! This, surely marks the occasion when the Milankovitch model came in out of the cold to be welcomed by the meteorological establishment.

Mason's approach to the problem was so simple that today the surprise is that no one had thought of it before. But it does depend upon a good knowledge of how ice sheets have advanced and retreated over the past 100,000 years, and this was only established in the previous decade or so. The calculations involved need no high-speed electronic computers, but can be performed literally 'on the back of an envelope'—the physicists' usually mythical criterion of absolute simplicity.

In order to melt ice, heat must be supplied to overcome the 'latent heat of fusion', at a rate of 80 calories for every gram of ice melted. In other words, to turn a given amount of ice at 0°C to water at 0°C requires as much heat as raising the temperature of that water from 0°C to 80°C. And when you are melting glaciers, that adds up to a lot of heat. This is because water is a more energetic state of matter than ice—and the vapour is more energetic still than the liquid. The 'latent heat of evaporation' is much greater than the 'latent heat of fusion'.

Water vapour can exist even at a temperature of 0°C, along with its liquid and solid phases. But when the vapour condenses into water at the same temperature, almost 600 calories of heat is released. If the vapour goes all the way to the solid and falls as snow, about 675 calories of heat is released in all from each gram. This heat goes into warming the surrounding air and spreads around the globe to affect worldwide temperatures. Equally, the need for heat to be absorbed in melting snow and ice helps keep cool regions covered by winter snow well into the early summer. Only after the snow is melted can summer heat up the land and air near the ground. Each year the cycle of the seasons shows us in miniature the pattern of an ice age. Ice cannot build up as soon as the temperature drops below 0°C, but only when, and if, the total deficit of the heat budget is sufficient to overcome the latent heat barrier. Equally, snow or ice cannot disappear as soon as temperatures rise above freezing, but only slowly as the heat budget builds up to push past the barrier in the other direction.

Dr Mason made the appropriate calculations, using both Milankovitch's original estimates of insolation changes and more recent refinements of them by Professor A. D. Vernekar in the USA. Between 83,000 and 18,000 years ago, during the most recent glacial period, there was an overall deficit in insolation at high northern latitudes (compared with the present day) of no less than 45,000,000,000,000,000,000,000,000 (45 septillion) calories, equivalent to about 1,000 calories for each gram of ice that we know, from the geological record, to have formed over that time to produce great ice sheets covering much of the northern hemisphere. This is very close to the figure of 675 calories per gram needed to account for the heat loss associated with water vapour turning to snow, and allows a little margin for cooling the water vapour to 0°C in the first place.

From about 18,000 years ago to the present, all this accumulated ice has melted except for a small residue near the pole. During this time, the overall excess in insolation at high latitudes—the surplus on the heat budget—was about 4.2 septillion calories. And the amount of heat needed to melt the ice that has disappeared, at 80 calories a gram, is just 3.2 septillion calories! Even though the calculations are so simple and approximate, the agreement is far too good to be explained by coincidence. In dealing with these astronomical numbers, the difference between the 4.2 and 3.2 matters much less than the agreement on the number of zeros; 1 septillion is 1 million million million million, and any 'coincidence' which stretches so accurately across 24 orders of magnitude cannot be dismissed lightly.

So we can now accept the workings of the Milankovitch mechanism as real and use them to predict the future of climate on Earth. About 140,000 years ago increasing northern summer insolation brought the rapid end of the most recent ice age but one. Some 83,000 years ago a switch back into deficit on the heat budget brought a strong build-up of ice, held back, but not stopped by, a slight warming 50,000 years ago. This latest ice age ended about 18,000 years ago when a surplus developed on the heat budget and brought the present interglacial. Already this interglacial age is past its peak, and we are slipping over into the next full ice age.

There are other influences on climate—both very long-term variations, such as those caused by continental drift, and shorter-term ripples, such as the veiling effects of volcanic dust thrown high into the stratosphere. But with the continents arranged as they are today, the best way to look at the puzzle of climate turns out not to be to ask why there are ice ages, but rather why there is ever a temporary respite from the ice. Interglacials occur only when the changing patterns of the Milankovitch model are just right. The unusual feature is the interglacial state, which has persisted now for more than 10,000 years and which we would like to persist for a little while yet. The escape from ice age conditions, beginning 18,000 years ago, required the combined effects of all three Milankovitch cycles to drag the Earth into a peak of warmth (by recent standards) about 6,000 years ago. Low orbital eccentricity coincided with a shift in the wobble which made June the month of closest approach to the Sun, boosting the heat of northern summers. At the same time the roll reached a maximum, putting the summer Sun high in the sky and again boosting warmth at the critical time of year. But since 6,000 years ago, all these factors have turned around, and conditions are becoming steadily less favourable. The grim prospect is 60,000 years of unfavourable orbital geometry ahead, pushing us back into full ice age conditions.

But the situation is not quite as black as it might be. The next dip in the northern-hemisphere insolation curve is not likely to be as deep as any of the past three dips, and if past evidence is anything to go by, it could take two or three of the short cycles for the glaciers to build up—ample time, surely, for the human race to devise countermeasures, if we can survive the more immediate problems that we have brought upon ourselves. What the curves do suggest, very strongly, is that as well as a full ice age at most 10,000 or so years ahead, we are entering a phase like that from 120,000 to 100,000 years ago, when something between a full ice age and a genuine interglacial developed, with repeated fluctuations of climate about the harsh conditions of a little ice age (the grim period from about AD 1500 to 1850) or worse.

The prospect of a return to a full ice age is genuine, but remote, on the human timescale. But as a backdrop to this picture we can see the Earth sliding into the next ice age through a succession of mini ice ages, with each recovery failing to reach the previous peak and each trough a little deeper than the one before. This prospect is one that should be borne in mind, while waiting for the next ice age, by everyone concerned about the human race and its future as the owner-occupier of one rather small planet, growing frostier.

four

How Darwin Discovered Relativity

▼

THE HEAT OF THE noonday Sun in summer in Los Angeles would be able to melt a layer of ice 1 centimetre thick and 1 metre square in 52 minutes. This only begins to seem truly impressive when you realize that as far as we know the Sun is pouring out radiation equally energetically in all directions. So if there is enough energy to melt a piece of ice 1 centimetre thick that quickly at the point where the Earth in its orbit happens to intercept that solar radiation, there must also be the same amount of energy crossing every square metre of space at the same distance from the Sun. In other words, there is enough energy pouring out of the Sun to melt a whole shell of ice, one cm thick and 300 million kilometres in diameter (the diameter of the Earth's orbit) in 52 minutes—*every* 52 minutes.

Now, imagine this shell of ice shrinking in diameter, closing in on the Sun so that its area gets less and less, but with its thickness increasing so that it always contains the same amount of ice. By the time the inner surface of this shell touches the surface of the Sun, the imaginary layer of ice would be more than a mile thick—but it would still be thawed in the same brief span of time.

The temperature at the surface of the Sun, sufficient to achieve this feat, is 5,770 K (just over 6,000°C). We can measure this today by measuring the amount of heat arriving at each square foot of the Earth's surface or by warming the detectors of a satellite in space and making allowance for the distance to the Sun. The temperature can also be inferred in another way, from the colour of the Sun. Just as a white-hot piece of iron is hotter than a red-hot lump, so a blue or white star is hotter than a yellow or orange star. The colour—temperature connection follows a precise law, studied in detail in laboratory experiments. This shows that the temperature of a yellowish star like our Sun is indeed 5,770 K.

This is not, in truth, a particularly remarkable figure. Even the glowing filament of an electric light bulb runs at about 2,000 K; and although it is a little hotter than red-hot iron, the *surface* of the Sun, at least, is not at a temperature we have any real trouble comprehending. But how long has the Sun been this hot?

In the nineteenth century, when geologists and evolutionary biologists first began to appreciate the extreme age of the Earth, and pointed out that the Sun must have been shining steadily for many hundreds of millions, perhaps many thousands of millions of years, they started a scientific controversy that raged for decades. The problem was that no source of energy known to nineteenth-century science could possibly keep the Sun so hot for so long. Were the geologists and biologists wrong? Or was the understanding of physics deficient? One of the greatest scientists of the day was convinced that if something had to give, it certainly was not going to be the laws of physics, and he mounted a determined attack on anyone who dared to suggest otherwise. And yet, the weight of geological evidence certainly had to be taken seriously.

As recently as the eighteenth century, it was widely accepted that the Earth was created about 6,000 years ago. The estimate was made by counting back the generations listed in the Bible, from Jesus Christ to Adam. Today, even theologians accept that the Bible should not be taken literally at that level and that the Earth, our Sun and the Universe at large have been around for a vastly greater span of time than our recent ancestors could have imagined. The first attempt to extend the timescale—the first real scientific estimate of the age of the Earth—pushed this out only to 75,000 years or so, still far short of the figure calculated today. But that still increased the age estimate tenfold, and it flew in the face of established religious doctrine.

Georges-Louis Leclerc, Comte de Buffon, was the man who made the new calculation. He was born in 1707 at Montbard, in Burgundy, and he introduced science into the calculation of the age of the Earth. Buffon was not convinced that the heat of the Sun was enough to keep the Earth warm, and he assumed that heat from inside the Earth was essential to provide conditions suitable for life. Since he knew of no way in which heat could be generated inside the Earth today, he also assumed that the Earth had been formed as a molten ball of rock and had been cooling ever since. The molten Earth, he suggested, had been torn out of the Sun by a passing comet; but how long would it have taken to cool to its present state?

Buffon actually carried out experiments with balls of iron and other substances of different sizes, observing how long it took for them to cool down from red heat. Armed with this information, and the accurate knowledge astronomers already had of the size of the Earth, he calculated that if the Earth had been born in a molten state, it would have taken

36,000 years to cool to the point where life could appear, and a further 39,000 years to cool to its present temperature.

Theologians of the day attacked this extension of the timescale of Earth history. But Buffon's ideas had a lasting influence in the nineteenth century. The direct line of that influence on subsequent generations of scientists was through Joseph Fourier, who is best remembered today for his development of the mathematical tool known as Fourier series, later developed as Fourier (or harmonic) analysis. In fact Fourier, who was born in 1768 in Auxerre, was primarily a physicist and developed his mathematics as a means to an end, in order to be able to analyse accurately interesting physical problems. He was fascinated by the problem of providing an accurate means to calculate the way in which heat is transferred through an object. Publicity for Buffon's calculation of the age of the Earth led Fourier to his own study of heat conduction, and to the mathematics he needed to describe the process.

Buffon had simply measured the rate at which hot objects cooled and tried to extrapolate this up to an object the size of the Earth. Fourier developed mathematical equations to describe the rate at which heat is lost from a body and used them to calculate how long it would have taken the Earth to cool. On this picture, the important difference is that the Earth must be coolest on the outside but still at the temperature of molten rock at its centre (actually slightly hotter than the surface of the Sun), even today. There is a steady fall in temperature—a thermal gradient—from the inside to the outside, and a steady flow of heat outward. Because the layers of cooler material surrounding the hot core act as an insulating blanket, holding heat in, it takes much longer for the Earth to cool than Buffon had realized. In 1820, Fourier wrote down the formula for the age of the Earth, based on these arguments, but as far as anyone has been able to find out, he never wrote down the number that comes out of this formula. Perhaps he regarded the value he had derived as too big to be taken seriously—for, instead of Buffon's 75 *thousand* years, Fourier's equations implied an age for the Earth of 100 *million* years.

The figure caused no immediate stir, simply because it was not publicized. Fourier died in 1830, and it was to be another 30 years before essentially the same calculation was taken up and promoted widely as indicating the true age of the Earth. But by then, so rapid had been the pace of change that the enthusiastic promoter of this timescale, William Thomson, later known as Lord Kelvin, was making the case that 100 million years was such a *short* timescale that it ought to be causing embarrassment to geologists and evolutionists!

It was, indeed, from geology and the idea of evolution that nineteenth-century science was driven to consider such long timescales for the history of the Earth and, by implication, the Sun. The idea that the Earth had a much longer history than even a few million years was propounded in the nineteenth century by a Scot, Charles Lyell, who was born in 1797. In the late 1820s, Lyell travelled extensively on the European continent, and everywhere he went he saw evidence of the way natural forces could mould the features of the Earth—the region around Mount Etna, with volcanic activity at work, made a particular impact on him. The fruits of Lyell's travels appeared in his three-volume work *Principles of Geology*, published in the 1830s. He promoted the idea of uniformitarianism—that the same processes we see at work today (such as volcanic activity and erosion), operating over very long periods of time, are sufficient to explain all the features of our planet, including mountain ranges and river valleys.

Lyell's book caused an immediate stir. It made a particularly striking impression on a young naturalist just setting off on a voyage in HMS *Beagle*. Charles Darwin never failed to acknowledge his debt to Lyell, who showed him that the Earth was very old indeed and that all that was needed to explain its present appearance was the same set of forces that we see at work today. Lyell applied that doctrine to the rocks; Darwin applied it, with equal success, to living things. Evolution by natural selection requires, above all else, a long timescale in which to operate. Lyell gave Darwin that timescale.

Both Darwin and Lyell came under attack, not only from religious fundamentalists but also from physicists who argued that no known natural processes could have provided conditions suitable for life on Earth for long enough for geological processes to have shaped the planet, or for evolution to have produced the diversity of life we see today. There was no obvious answer to these criticisms, which had to be taken seriously. Biology and geology seemed to be telling scientists that the Earth and the Sun were much older than was physically possible.

The impossibility, according to nineteenth-century physics, of the Sun being able to shine for hundreds of millions of years emerged from one of the greatest achievements of science: thermodynamics. The realization that heat energy is exactly equivalent to mechanical energy (work), that heat flows only from a hotter object to a cooler one, and that the amount of disorder in the Universe (entropy) is always increasing, revolutionized science and made it possible for physicists to investigate and quantify many phenomena that had previously been inexplicable in strictly scientific terms. One of those phenomena to be investigated—by no means regarded at the time as the most important piece of nature to be caught in the thermodynamic net—was the age of the Earth and Sun.

In those days, scientists who began to worry about the profligate way the Sun pours energy out into space naturally thought in terms of coal, the fuel that powered the Industrial Revolution. If the Sun were made entirely of coal, burning in the most efficient way possible, then it could maintain its heat for only a few tens of thousands of years. No other form of chemical energy supply could extend this lifetime very much. But a few nineteenth-century physicists eventually realized that if chemical energy is insufficient to keep the Sun hot for more than a few tens of thousands of years, the only other source of energy they knew of could extend the timescale enormously. That energy source is gravity.

In terms of the conservation of energy (one of the key features of thermodynamics), what they needed was some reservoir of energy that could be drawn upon steadily for millions of years and converted into heat. Gravity might fit the bill, if a way of converting gravitational energy into heat could be found.

The first suggestion was that the Sun might be kept hot if it were 'fuelled' by a continuous supply of meteoroids falling on it from space. This is a source of energy which comes directly from the gravitational field of the Sun. When a meteoroid—essentially a lump of rock in space—falls towards the Sun, it does so because of the mutual gravitational attraction—or force—between the Sun and the meteoroid. Gravitational energy is converted into kinetic energy (energy of motion) as the meteoroid falls faster and faster. When the fast-moving rock hits the 'surface' of the Sun and stops, all that energy has to go somewhere. In exactly the same way, when a fast-moving car is brought to a halt using its brakes, all the energy of motion of the car has to go somewhere. The car's energy is converted into heat in the brakes, and can easily be felt if you hold your hand near the brake disks or drums just after the vehicle has stopped. In the case of meteoroids falling into the Sun (or, indeed, onto the Earth), the kinetic energy is also converted into heat, raising the temperature of both the meteoroid and whatever it happens to hit. When a meteorite strikes the Earth, the impact can melt rock explosively, blasting out a huge crater with the power of many millions of tonnes of TNT—far bigger, in extreme cases, than any artificial explosion, including nuclear blasts.

Because the Sun contains more mass than the Earth, it has a correspondingly stronger gravitational field, and meteoroids fall that much faster when they hit the Sun. The energy released is even greater than it would be if the same meteoroid hit the Earth. In principle, you could indeed make the Sun hot in this way, if there were enough meteoroids around to fall into it. It is not against the laws of physics. But there are nowhere near enough meteoroids around to do the job. The man who showed this, and found another way to keep the Sun hot, was William Thomson, who came across the meteoroid impact theory at the annual meeting of the British Association in 1853. He was immediately taken with the idea, and set out to calculate how long the Sun could be kept hot by such means. But although he spent a lot of time trying to make the idea work, eventually he had to admit defeat.

There is no need to go through all the painful steps of Thomson's attack on the problem, since the ultimate version of the theory makes its deficiencies brutally plain. As it became clear that there were not enough small, rocky objects in the Solar System to provide the required input of energy to the Sun, Thomson toyed with the idea that the Sun might maintain its fires by consuming not meteoroids but whole planets, one by one. In this picture, Mercury, the innermost planet, might slowly spiral into the Sun, giving up its gravitational energy as heat—but this would be sufficient to keep the Sun hot for only seven years. Venus would do little better, providing enough energy to heat the Sun for 84 years, and even Neptune, the most distant large planet in the known Solar System, could contribute only enough energy to keep the solar fires hot for some 2,000 years. Even by gobbling up all the planets in turn, the Sun still cannot maintain its fires for more than a few tens of thousands of years—the 'meteoritic' fuel supply is no better than the chemical one.

By the 1860s, Thomson was investigating another idea, but still involving gravity. By then, however, he had been pre-empted by a German researcher, Hermann von Helmholtz. Helmholtz was born in Potsdam in 1821. His major contribution to the debate about the origin of the Sun's energy appeared in February, 1854. The brilliantly simple idea that Helmholtz came up with was the suggestion that the whole mass of the Sun itself might provide the gravitational energy to make it hot.

The argument is straightforward. If the whole Sun were made of rock and all the pieces were carried far out into space, then each piece would have a large amount of gravitational energy, and they would all fall towards the centre of the cloud of stones. The gravitational energy provided by the whole mass of the Sun itself, initially spread out in a cloud of rocks then falling inward (converting gravitational energy into kinetic energy) and bashing together in a molten ball of fire (converting kinetic energy into heat) would release as much energy as the Sun radiates in 20 *million* years.

In 1854, Helmholtz had not made the precise calculation and simply pointed out that a great deal of heat could be produced in this way. Thomson put the numbers in, but at first he didn't think much of the proposal—he considered the idea that the original material of the Sun was a dispersed cloud of stones to be rather implausible. Besides, what was the point of having 20 million times as much energy as the Sun radiates in a year produced all at once, in a great explosion? What was needed was a way to release at least as much energy, but *slowly*, over millions of years. It wasn't until the end of 1860 that Thomson realized that the idea put forward by Helmholtz actually did provide just that possibility.

His new insight was all thanks to a happy accident (happy for science, painful for Thomson) which left him with a broken leg and ample time to lie in bed thinking. This was just a year after the publication of Darwin's *The Origin of Species*, and that may well have been one of the reasons why Thomson thought about the origin of the Sun's energy supply and the problem of the age of the Sun and Earth. The fruits of his thinking appeared in *Macmillan's Magazine* in 1862, and made a big impact.

At first, Thomson based his arguments very much on the image of a mass of stony meteoroids coming together, distasteful though that had seemed to him earlier. He didn't worry too much about how the vast amount of energy available was stored up and allowed to trickle out slowly over millions of years, but simply about how much energy was available. In round terms, the theory showed that there was enough energy to keep the Sun shining at its present rate for between 10 and 20 million years. Even allowing for the possibility of errors in the calculation, or in the assumptions on which it was based, Thomson could see no way in which this might be increased by more than a factor of 10 and wrote,

> It seems, therefore, on the whole most probable that the Sun has not illuminated the Earth for 100,000,000 years, and almost certain that he has not done so for 500,000,000 years. As for the future, we may say, with equal certainly, that inhabitants of the Earth cannot continue to enjoy the light and heat essential to their life, for many million years longer.

Thomson's calculation, flatly contradicting the timescale needed by Darwin for evolution to do its work, initiated a debate which continued for the rest of the nineteenth century. As a result of that debate, Thomson continued to refine his calculations, and it was not until 1887 that he came up with the final version of the estimate of how much energy could be provided to keep the Sun hot by gravitational collapse.

The idea actually drew upon a suggestion made by Helmholtz in his first paper on the Sun's heat, back in 1854. But when Thomson presented his own calculations in a lecture at the Royal Institution in London in 1887, he made no mention of Helmholtz and may well have forgotten the details of the paper he had read 33 years before.

The important feature of this final step in Thomson's work on the heat of the Sun was the realization that as long as the same amount of matter—the same mass—is involved, it is not important how big, or how small, the original 'rocks' from which the Sun formed might be. Just two half-Suns, falling directly onto each other from far away, would collide with as much kinetic energy as a cloud of pebbles collapsing to its centre. And so would a cloud of hydrogen atoms, the lightest 'particles' known to nineteenth-century science. There is just as much energy available if the original cloud from which the Sun formed was made of gas spread out initially over a huge volume and collapsing under its own weight. By the time such a collapsing gas cloud had shrunk to roughly the size the Sun is today, the temperature at its core would be millions of degrees, and its surface would glow with a temperature of a few thousand degrees. Astronomers today accept this broad picture as the most likely explanation of how stars form in the first place.

Once such a proto-star is hot inside, there is a great deal of pressure pushing outward, because the heat makes the particles of the gas very energetic. They jostle vigorously against one another. This pressure holds the star up and stops it from shrinking. As long as the star is hot inside, it can never collapse completely. But what happens if the ball of gas cools down slightly and the pressure drops? Obviously, it will shrink. Thomson realized that what shrinking really means is all of the particles in the Sun moving a little closer to the centre—*falling* in a gravitational field. So they would gain kinetic energy, which is converted into heat when they jostle against one another. The pressure would increase again and resist the collapse. All that was required to ensure that the gravitational energy stored in the Sun was released slowly, over millions of years, was that the Sun should be shrinking slowly. Thomson calculated that if the Sun shrank by just 50 metres a year, that would liberate enough energy to keep it as hot as we see it—hot enough to melt a shell of ice around it, something like 2 kilometres thick, every 52 minutes. There could still be no more than enough energy to last for about 20 million years, but at last he had found a way to dole that energy out year by year, instead of releasing it all at once.

In 1892, the year he received his peerage at the age of sixty-eight, the new Lord Kelvin stated a conclusion based on forty years' work:

> Within a finite period of time past the Earth must have been, and within a finite period of time to come must again be, unfit for the habitation of man as at present constituted, *unless operations have been and are to be performed which are impossible under the laws governing the known operations going on at present in the material world.*

And by 1897 Kelvin had accepted that the best estimate for the age of the Sun and Earth was 24 million years, a figure known today as the Kelvin—Helmholtz (in Germany, the Helmholtz—Kelvin) timescale.

All of the calculations were impeccable. There is, indeed, no way that a star like the Sun can keep itself hot for more than about 20 million years by slow contraction. And yet, well before the end of the nineteenth century, it was clear that this figure was far too small to meet the requirements of geology and evolution. Darwin's theory, in particular, insisted that the Sun was older than it could possibly be, according to nineteenth-century physics. Something had to give. We now know that the something extra required by evolution to keep the Sun hot is the conversion of mass into energy, in line with Einstein's Theory of Relativity.

The way ahead was pointed most clearly in 1899, neatly at the end of the century, by Thomas Chamberlin, Professor of Geology at the University of Chicago. Writing in the journal *Science*, he commented:

> Is present knowledge relative to the behavior of matter under such extraordinary conditions as obtained in the interior of the Sun sufficiently exhaustive to warrant the assertion that no unrecognized sources of heat reside there? What the internal constitution of the atoms may be is yet open to question. It is not improbable that they are complex organizations and seats of enormous energies.

'Seats of enormous energies', indeed. The scientific community was ready for a completely new explanation of how the Sun maintained its fires. Evolution, together with the strict limit of the Kelvin—Helmholtz timescale, had shown the need for nuclear energy, even before it was discovered here on Earth. But that, of course, is another story.

five

How 'Normal' Is Our Sun?

▼

THE SUN IS THE nearest star to us, dominating the Solar System and, it might naturally be thought, providing us with the archetypal 'normal' star against which astrophysicists can conveniently check their theories of stellar structure. Until the 1970s, the Sun's track record as an archetype was pretty good, and its mere existence over the thousands of millions of years of the geological record had been enough to show that the fusion of light elements into heavier ones—nuclear burning—must be the power source for the stars. Throughout the 1950s in particular, the astrophysical implications of this nuclear burning process were developed to the point where the theory explained the existence of the variety of stars seen in the Milky Way, including exotically named varieties such as white dwarfs and red giants, while nova and supernova explosions could be seen as the result of runaway nuclear burning under critical conditions (although it must be admitted that the details of supernovae, in particular, remain to be worked out even today).

In the late 1960s and early 1970s, however, astrophysicists were faced with a series of puzzles, not about the exotica of the Milky Way but about the Sun itself. There are two ways of looking at these puzzles. Perhaps it's just that the Sun is so close that the 'warts' show up to spoil the beautiful agreement between theory and observation which seems to be there when we view stars from farther away. Or perhaps that agreement is genuine when we are dealing with stars in general, but no longer holds true in detail for the special case of our own Sun. Either way, the situation is embarrassing; but before we can see just how embarrassing we need some idea of just what the standard theories are.

Inside a star the temperature reaches many millions of degrees, initially through the heat produced as the gravitational energy of a collapsing gas cloud is liberated. At these temperatures, atoms are competely stripped of their electrons so that the stellar material is a plasma—a sea of atomic nuclei (positively charged) mixed in with negatively charged electrons. As far as the nuclear fusion processes that maintain these temperatures after the initial collapse are concerned, we can ignore the electrons and briefly sketch in a rather oversimplified picture of at least the first steps on the nuclear fusion ladder.

Most of the nuclei are those of hydrogen—individual protons—which naturally tend to repel one another since they all carry the same positive electric charge. If, however, the plasma is hot enough, some of the protons have enough kinetic energy to overcome this repulsion during collisions and can then combine in pairs. Each pair of protons that sticks together ejects one unit of positive charge (a positron or anti-electron, which soon meets an electron in the plasma and is annihilated), leaving behind one proton and one neutron, or a nucleus of deuterium (heavy hydrogen). Deuterium can then combine with another proton to produce helium-3 (2 protons, 1 neutron), and two nuclei of helium-3 can get together to produce one nucleus of helium-4 (2 protons, 2 neutrons) plus two ejected protons which go back into the pot. The net effect is that four protons have been converted into one nucleus of helium-4, and along the way a little mass (about 1 per cent of the mass of four protons) has been converted into heat in line with Einstein's notorious equation $E = mc^2$. The energy being produced by the sun today is equivalent to the destruction of 4.5 million tonnes of mass every second, a mere flea-bite compared with the total mass of the Sun, which is 2000 million million million million tonnes.

The nuclei of other elements are also built up in stars through chains of nuclear fusion reactions. More energy (which means a higher temperature) is needed for fusing more massive nuclei, and a star like the Sun is, according to theory, almost entirely powered by hydrogen burning. But a few unusually energetic nuclei are always around, and even in the Sun occasional reactions build up elements more complex than helium—and 'occasional' soon mounts up by terrestrial standards when such a large mass of material is involved. This is why the above picture is an oversimplification.

Still, it seems worth sketching the later stages of evolution of a star in an equally simple way. As hydrogen burning continues, the centre of the star eventually becomes clogged with an 'ash' of helium nuclei, so that eventually the heat generated is no longer sufficient to hold the star up against gravity. When that point arrives, the core of the star collapses, heats up and, with more energy to play with, helium nuclei start to fuse together in large numbers. In this helium burning stage the extra heat from the compact core forces the outer layers of the star to expand, and it becomes a giant. Meanwhile, the fusing of more massive nuclei begins to build up more interesting elements (as far as we are concerned) including carbon and oxygen. These elements are dispersed across the Galaxy when stars explode at the end of the stage of stable nuclear burning, eventually finding a place in the clouds of interstellar material from which new stellar and planetary systems form, planets on which living creatures composed chiefly of 'heavy' elements (like ourselves) can evolve.

The real evidence that all these nuclear processes go on inside stars came with the success of computer models in which numerical simulations of the various nuclear processes, together with a specification of the mass of a star, are sufficient to produce a 'prediction' that a star of a certain mass and age will have a certain surface temperature (or colour) and composition. Such models 'predict' that a star of 1 solar mass and the same age as our Sun will be as bright as the Sun and have the size of our Sun—or at least, very nearly. And it is the gap between 'exactly' and 'very nearly' that leads us now to question the 'normality' of our Sun.

Thirty years ago, it seemed that only details of stellar nucleosynthesis remained to be worked out and that we understood the workings of an ordinary star like the Sun pretty thoroughly. But this view was wrong, and it was shown to be wrong when it became possible, armed with the best theories, to make direct 'observations' of something happening in the interior of the Sun. But that 'something' turns out not to be there—and that discovery (or lack of discovery) is what upset the theorists' applecart.

Conventional observations of the Sun and stars depend on monitoring the electromagnetic radiation from their surfaces, but a photon struggling out from the interior can take millions of years to reach the surface, being involved in many collisions and interactions on the way. Conventional astronomy tells us only about the average conditions inside the star—an average effectively taken over this very long timescale for radiation to travel through the hot plasma. To the great pioneers of the theory of stellar structure, such as Arthur Eddington, the concept of a particle that can travel from the centre of the Sun out and across space to reach the Earth in only a few minutes would have seemed almost magical, but those pioneers would certainly have appreciated the value of such a particle to anyone wishing to study the reactions going on in the solar interior today. Just such a particle—the neutrino—is now an essential feature of our best theories of nuclear physics, and evidence for the existence of neutrinos has come from many experiments involving particle interactions here on Earth. According to these theories, many nuclear reactions just cannot work without neutrinos being involved—indeed, it's just because particular reactions do occur that neutrinos are believed to exist. But the neutrino is certainly an odd bird. It has no mass, travels at the speed of light and carries 'spin'. It is so unsociable that only one neutrino in every hundred thousand million produced in the Sun should, according to theory, fail to escape into space. To a neutrino, the fiery Sun and solid Earth are scarcely more of an obstruction than empty space itself. But elusive though they may be, according to those same best theories 3 per cent of the energy radiated by the Sun is in the form of neutrinos, and just about a hundred thousand million solar neutrinos cross every square centimetre of the Earth (including you, me and this page) every second. That is so many that we ought to be able to catch some with suitable detectors, even though they are so reluctant to interact with anything at all. It would be hopeless to try to catch one solitary neutrino, but reasonable enough to hope to trap a few out of that hundred thousand million crossing each square centimetre each second.

In the attempt to do just that, researchers have built detectors as large as a swimming pool and then buried them deep in a mine underground, where the influence of everything except neutrinos is screened out by layers of rock. The detectors are filled with carbon tetrachloride, for neutrinos are believed to be partial to some of the nuclei of the chlorine in that compound. In theory, a few solar neutrinos should interact with the isotope chlorine-37, with the end result that a few atoms of argon-37 will be produced in the tank of fluid, and argon-37 is much easier to detect than a neutrino. The snag is that only about one-third of the expected number of argon-37 atoms turn up in the tank. This has now been going on for thirty years, and in the past ten years other experiments around the world, using slightly different techniques, have confirmed the deficit in the number of solar neutrinos reaching the Earth. Only two possibilities are allowed, assuming the observations are correct. Either the theories of particle physics are wrong, or the theories of stellar structure are wrong, at least when applied to the Sun.

Now, the particle interactions can be tested to some extent using accelerators on Earth—indeed, it's the results of those experiments that provide the basis of the theory that suggests the Sun should produce a lot of neutrinos. It is very difficult to see how the nuclear and particle physics theories could be adjusted slightly to explain away the lack of solar neutrinos detected on Earth. Several theorists have tried more or less bizarre adjustments to the theory, but none of them really work. On the other hand, it is very easy to adjust the structure of a computer model of a 1 solar mass star to switch off the flow of neutrinos. All you need to do is turn down the temperature of the nuclear cooker slightly—by about 10 per cent—which is enough to turn off the nuclear reactions that are expected to produce most of the neutrinos. Of course, that can only be a temporary solution since the Sun needs those reactions to keep it hot.

But is it possible that the Sun has actually gone 'off the boil' temporarily? Remember, if this idea is correct, the lack of neutrinos indicates that the Sun's interior is a little cooler right now than it is on average; the surface brightness, on the other hand, merely confirms the temperature of the interior averaged over many millions of years. If nuclear fusion in the interior were switched off entirely, the Sun would not go out like a light but would settle down slowly over about 30 million years. The geological and other records show that the Sun has been burning fairly steadily, thanks to nuclear fusion, for thousands of millions of years, but the inbuilt safety valve provided by its store of gravitational potential energy means that it can continue superficially unchanged for several million years if something does happen to disturb the normal nuclear burning. So, on the face of things, the neutrino evidence suggests that the Sun is not in a 'normal' state just now. Not normal, that is, when compared either with its own long history or with the present state of most stars in the Milky Way. The problem then becomes one of explaining how and why the Sun might have gone off the boil temporarily, and it turns out that an intriguingly plausible possibility has been around for more than 25 years, associated with a theory of ice ages. As an explanation of ice ages the theory doesn't really look all that convincing, but ideas about intermittent convective mixing in the Sun seem highly appropriate in the context of the modern solar neutrino problem.

The basis for the model is that from time to time conditions inside the Sun may become appropriate for convective mixing to occur over a large part of the interior, producing first an increase in heat and an associated slight expansion of the outer layers as more hydrogen fuel is mixed into the nuclear burning region. Then a phase of cooling and contraction takes place as conditions return towards the long-term equilibrium. The idea was picked up by several theorists, including William Fowler, and even linked with theories of climatic change on Mars. But the snag with all these ideas is that no one actually explained why the convective mixing should just happen to have occurred in the recent past ('recent' on stellar evolution timescales, that is). The problem is still there, but one step further removed: no solar neutrinos because of recent mixing, perhaps, but why recent mixing?

Late in 1975 a possible answer appeared, again in a theory of ice ages and curiously with no mention at the time of the possible implications for solar neutrinos. Bill McCrea presented his model of how the interaction of the Sun with the dark lanes of compressed material which define the spiral arms of our Galaxy could produce ice ages, comets and, originally, the Solar System itself. That is quite a story in itself, but it is McCrea's to tell, not mine; the important point for the neutrino puzzle is that the Sun can be affected by the dust it gathers in while passing through such a spiral feature, and that we have just passed through the dust lane associated with the Orion Arm (indeed, we may even have passed through the dense material of the Orion Nebula itself, emerging less than 20,000 years ago). If the dust accreting onto the Sun can trigger the convective instability, we are home and dry; a disturbance only 20,000 years old would certainly not yet have been worked through the Sun's system and equilibrium been restored. But how can we test the idea?

Once again, we come back to the computer models. The Sun gathering in material from a cloud of dust is rather like a star in a binary system gathering in material from its neighbour, and many people have run computer simulations of that situation. What they show is that the energy from the infalling material causes alterations in the outer layers of the star (of course) with a very slight decrease in the rate of nuclear burning, and a tendency to stabilize the star against convective instability. That seems just the opposite of the effect we are looking for, but what happens when the extra material stops falling in—when the Sun emerges from the cloud of dust? This is much harder to answer, since the computer models have not yet been able to produce meaningful answers when asked that question. In computer jargon, the models 'fail to converge'. That in itself could be taken as indicating the possibility of instabilities setting in. Computer models find it hard to cope with sudden changes. And in physical terms, if adding material inhibits convection, suddenly ceasing to add material could certainly encourage the spread of convection, in a way very similar to the effects of removing the lid from a pressure cooker. Such evidence as there is—and admittedly it is not conclusive—suggests that *emergence* from a dust cloud could indeed trigger convective instability, allowing the temperature of the solar interior to drop temporarily by the 10 per cent or so needed to stop the production of large quantities of neutrinos. To some extent the evidence is circumstantial, but since McCrea's ideas were published yet more circumstantial evidence pointing the same way has also appeared.

One clue comes, yet again, from climatic studies. Researchers at the National Center for Atmospheric Research in Boulder, Colorado, have looked again at the old idea that sunspots are associated with changes in the weather and that climate varies in line with the changing level of solar activity. By including a variation of the amount of heat radiated by the Sun (the solar 'constant', which they now dub the solar 'parameter') these researchers can explain many features, of changing global temperatures over the past three centuries. The variation they allow seems remarkable to an astronomer brought up to think of the Sun as a steady, stable star—as much as 2 per cent over the 11-year cycle. Most remarkable of all, however, such observations of the solar parameter as there are certainly do not rule out the reality of such a variation.

Indeed, recent observations of the outer planets show

that their brightnesses vary over the roughly 11-year sunspot cycle by just this amount (2 per cent), although again no one seems to believe this is an effect of the changing solar parameter, but rather due to changes in the composition of the clouds in the atmospheres of those planets, brought about by photochemical effects linked with the changing solar cycle. Another coincidence?

The overall situation is tantalising. We certainly have enough evidence of one kind and another to make the question 'How normal is our Sun?' very well worth asking, but we do not at present have enough information for an unequivocal answer one way or the other. This is a vital question for physicists pushing our understanding of the laws of physics to the limit, for if the Sun is *not* 'normal', then their theories are on surer ground. Equally, as the link with ideas of climatic change (whether it be ice ages or the smaller fluctuations of the past three centuries) shows, this question is of far more widespread interest. We may not all be concerned about details of the laws of physics, but we are certainly all concerned about changes in the Sun which can affect the climate on Earth. It is also of more than passing interest to ponder on the evidence that the development of humans as intelligent adaptable creatures was influenced in no small measure by the advance and retreat of the ice a few tens of thousands of years ago. We may owe our very existence on Earth at the present time to the same astronomical factors that are responsible for the abnormal state of the Sun today. As McCrea commented, 'life can be very complicated'.

six

The Curious Case of the Shrinking Sun

▼

ASTRONOMERS WERE STARTLED, AND laymen amazed, when in 1979 Jack Eddy, of the High Altitude Observatory in Boulder, Colorado, claimed that the Sun was shrinking, and at such a rate that, if the decline did not reverse, our local star would disappear within a hundred thousand years. Working with mathematician Aram Boornazian on a study of measurements of the Sun's diameter made at England's Royal Greenwich Observatory between 1836 and 1953, Eddy found evidence of a decline in solar angular diameter of two seconds of arc—equivalent to one-tenth of 1 per cent—per century. Since we have very good evidence indeed that the Earth and the Sun have existed with essentially the same relationship to each other as at present for at least 4 thousand million years, such a discovery could only imply that the Sun is at present in a temporary phase of contraction, which must soon be halted and reversed.

Startling though these claims seemed to many people, to others they were almost welcome. For, throughout the

1970s, astronomers had been puzzling over the mystery of the missing solar neutrinos. A shrinking Sun could resolve the solar neutrino puzzle, because a shrinking Sun generates some of its heat by the release of gravitational potential energy. This reduces the amount of energy generation required from nuclear reactions in its heart to explain the observed surface luminosity. With a reduced output from the central powerhouse, fewer neutrinos would be produced, perhaps in line with the numbers actually detected by Ray Davis in his famous tank of cleaning fluid buried in a gold mine in South Dakota.

Alas for such hopes, Eddy's initial claims quickly turned out to be too much of a good thing. The observations at Greenwich on which the claims were based had been made by different observers at different times, using techniques which depended, to some degree, on the observers' skill and expertise in their chosen technique. One approach is to time the passage of the Sun, from limb to limb, across a fixed meridian wire, and calculate its size from the known rotation rate of the Earth; another depends on measurements of the solar disk using a micrometer screw at the eyepiece of a telescope. And both proved to have been subject to persistent human errors, which accounted for at least part of the apparent change in the size of the Sun.

Theorists pointed out that any such change in the size of the Sun ought to produce a detectable change in the measured energy output of the Sun—the solar constant—but that measurements of the solar constant over the period since 1850 implied that the solar radius must have been constant to within 0.3 seconds of arc per century—less than half the variation in the Greenwich records. And in mid-1980 Irwin Shapiro published an analysis of observations of the transits of Mercury across the Sun's disk, concluding that the Sun had maintained a constant size since the end of the seventeenth century. This technique depends on timing the passage of Mercury across the face of the Sun in those years when, in May and November, the Earth, Mercury and the Sun are appropriately aligned; this only happens about thirteen times each century, but such transits have now been the subject of detailed astronomical scrutiny for more than two hundred years.

By late 1980, the issue seemed clearcut. The old Royal Greenwich Observatory records contained a consistent error, and Eddy had been wrong to take them at face value. In November of that year, the *Griffith Observer* carried an article from Margaret L. Silbar with the title 'Is the Sun really shrinking?' and which left readers with little doubt that the answer was 'no.' But the plot began to thicken, and now the curious story of the shrinking Sun really begins. In mid-December 1980, both of the leading weekly science journals, *Nature* and *Science*, carried, in the same week, scientific papers reporting work stimulated by the claims of Eddy and Boornazian. As far as the numbers that came out of their calculations were concerned, the two papers reached exactly the same conclusions. Yet the interpretation of those numbers by the two teams involved led to diametrically opposed conclusions, as typified by the titles of the two papers. In *Science*, David Dunham, Sabatino Sofia, Alan Fiala, David Herald and Paul Muller presented their analysis under the heading 'Observation of a probable change in the solar radius between 1715 and 1979', while at the same time in *Nature* John Parkinson, Leslie Morrison and Richard Stephenson were proclaiming to the scientific world 'The constancy of the solar diameter over the past 250 years'. A close look at those two intriguing papers shows how even the most objective scientist can unwittingly colour his conclusions to suit his expectations, but leaves the question of solar size variations very much still open.

The solar constancy argument is essentially the one I have already outlined. Parkinson and his colleagues pointed out that, out of seven regular observers using the Royal Greenwich Observatory's meridian circle since 1851, five produced self-consistent observations of the Sun's diameter over each of their periods as observer, and the other two were full of 'strong, erratic personal biases'. Together with the Mercury transit data, this reinterpretation of the old records led to the conclusion that the maximum extent of any change in the Sun's radius since 1850 was no more than a decline of 0.08 seconds of arc per century, with a possible error range of plus or minus 0.07. Within the range of possible error, the Sun's diameter might be constant—the conclusion proclaimed in the *Nature* headline—or it might be shrinking, although at only one-tenth of the rate claimed by Eddy. It might not seem obvious which of the two conclusions to take on board, if it were not for the happy coincidence of the simultaneously published *Science* paper. For the conclusion reached by Dunham's team was that the Sun has indeed been shrinking and that its radius—or, strictly speaking, half the measured angular diameter—has contracted by 0.34 seconds of arc over 264 years—almost exactly in line with the decline of, in round terms, some 0.1 seconds of arc per century suggested by the re-examination of the Greenwich records!

The technique used by Dunham's team is disarmingly simple in concept, but it does depend upon the precise accuracy of some observations made in 1715 under the direction of Sir Edmond Halley, later Astronomer Royal and known to this day for the comet which bears his name. In that year, there was a total eclipse of the Sun visible from Britain on 3 May. Halley organized observations of the eclipse from different parts of the country; the duration of totality at each observing site provides an important clue to the size of the Moon's shadow on the Earth, and thereby to the size of the Sun at the time of the eclipse. An observer just a kilometre inside the edge of the path of totality, for example, would see a total eclipse lasting for only 15 seconds, and thanks to Halley's efforts valuable observations of this kind were made at both the northern and southern edges of the eclipse path in 1715.

In the north of England, Theophilus Shelton, Esquire, recorded at Darrington (a small town located just north of 53 degrees 40 minutes latitude, near the city of Leeds) that the Sun 'was reduced almost to a Point, which both in Colour and Size resembled the Planet Mars', concluding that the northern limit of totality was just south of his location. In the south of England, the edge of totality was bracketed by observations just either side of the line, south of the village of Cranbrook (which lies in the heart of Kent, just south of latitude 51 degrees 10 minutes), giving a precise guide to the extent of the Moon's shadow.

Now, the path of the edge of totality depends on the precise geometrical alignment of the Sun, Moon and Earth at the time, and this can be calculated very accurately from the standard equations of celestial mechanics. It also depends on the size of the Moon (but no one suggests that this has changed since 1715) and on the size of the Sun. Comparing Halley's data for the 1715 eclipse with similar observations of eclipses in 1976 (visible from Australia) and 1979 (watched by hundreds of amateur astronomers across North America), Dunham and his colleagues concluded that there was no measurable change in solar radius between 1976 and 1979, but that between 1715 and the 1970s the Sun had shrunk by 0.34 seconds of arc, with an uncertainty of plus or minus 0.02 seconds of arc. This disagrees with the earlier claim by Eddy and Boornazian, but falls within the limits set by theorists (including Sofia, himself a member of the group headed, alphabetically, by Dunham) on the basis of measurements of the solar constant. It is also consistent with the Mercury transit data, used by Parkinson's group as evidence for the constancy of the solar radius, but which actually say only that any change is smaller than 0.15 seconds of arc per century. To put all this in perspective, the Sun's angular diameter is, in round terms, 32 minutes of arc, just over half a degree; its linear diameter, just over 108 times that of the Earth, is 1,392,000 km, and it looks so small only because it is at a distance of just under 150 million km.

If those two papers had appeared out of the blue, with no preparation of the scientific ground, they would surely have made an enormous impact both among astronomers and in the wider world. For, of course, a shrinking of the Sun at a rate of merely 0.01 per cent per century still implies its total disappearance in a million years, or that it was twice its present size a million years ago (both ludicrous suggestions on the basis of everything that has been learned over the past half-century about how stars work), or that the present contraction is simply one phase of a long, slow pulsation which might have a cycle time of hundreds or thousands of years.

The snag was, all the headlines had already been written when Eddy and Boornazian came up with the suggestion that the Sun was shrinking at a rate of one-tenth of 1 per cent per century. Contraction at just one-tenth of this rate seemed small beer to astronomers, and was presented by the popular media (where they took any notice at all) as just another example of a way-out scientific idea that had had the rug pulled from under it by more careful studies. Metaphorically, we can imagine theoretical astronomers breathing a sigh of relief and saying 'Oh, so the Sun is only shrinking by a tenth of a second of arc per century, not a full arc second after all. Nothing to worry about.' It seems to have taken several months for the message to sink in that here was plenty to worry about, with deep implications for the inhabitants of planet Earth.

The immediate cause for concern is that changes in the

Sun's diameter are linked with changes in its heat output, and changes in the solar constant by even a fraction of 1 per cent can have a pronounced influence on the climate of the Earth. Eddy's own interest in the old records from the Royal Greenwich Observatory had developed from his study of the changing level of solar activity as indicated by the numbers of dark spots—sunspots—on its surface. He had proved during a thorough re-analysis of old observations that the coldest decades in recent history, the Little Ice Age of the second half of the seventeenth century, coincided (if that is the right word) with an interval when the Sun was remarkably free from these dark spots, although they have come and gone with a period of roughly 11 years ever since. Might changes in the size of the Sun be linked both to the surface activity manifested by sunspots and to small-scale climatic changes on Earth? Curiously, in their paper suggesting the constancy of the solar diameter, Parkinson and his colleagues did mention that the Mercury transit data show, as well as an overall decline in solar radius of less than 0.15 seconds of arc per century, hints of a periodic variation in the Sun's size with a cycle time of about 80 years. The special significance of this number is that a similar long-term cyclic pattern shows up in the record of the changing number of sunspots, modulating the stronger 11-year rhythm, and in climatic patterns revealed by historical records of temperature, the width of the annual growth rings in trees, and so on. Some powerful hints of what was to come were there to be seen by the knowledgeable in December 1980. But it was not until the autumn of 1981 that the bombshell finally exploded.

Hardly surprisingly, the breakthrough came from one of Eddy's colleagues, Ronald Gilliland, who also works at the High Altitude Observatory. Perhaps a little more surprisingly, it was published not in the pages of *Nature* or *Science*, where hot news is usually aired for rapid communication among scientists (and where, it has to be said, some of the publications prove on sober reflection to have been a little over-hasty), but between the sober covers of the *Astrophysical Journal*, a pillar of respectability among the astronomical establishment, and not a journal prone to giving space to half-baked ideas. Gilliland based his study on no less than five sets of data, including the old Royal Greenwich Observatory records with the correction for systematic observing errors, similar meridian circle measurements from the US Naval Observatory in Washington, DC, two sets of Mercury transit observations, and the solar eclipse data. His first conclusion, from a battery of statistical tests, was that the overall decline in solar diameter of about 0.1 seconds of arc per century since the early 1700s is real. And when standard statistical tests aimed at revealing small, regular changes in the pattern of variability were turned on the meridian circle data, they showed an unambiguously clear trace of a periodic variation with a repeating rhythm 76 years long—almost exactly matching up with the periodic variation present in the Mercury transit data.

This regular pulsation covers a range of only 0.02 per cent of the Sun's radius, but it has been stable over the full 250-year span covered by the various sets of observations, and it shows a clear relationship with sunspot activity with, by and large, fewer sunspots present when the Sun is bigger. If the same anti-correlation can be applied to the longer-term decline in the Sun's diameter, it may provide a clue to the dearth of sunspots during the height of the Little Ice Age, 300 years ago, when the Sun's angular diameter was about two-thirds of a second of arc greater than it is today. And as the icing on the cake, Gilliland also reported a smaller, but clearly present, fluctuation in solar size tying in with the 11-year sunspot cycle. The periodic variations are unambiguous. As for the longer-term decline in solar diameter, the discovery that started the whole ball rolling, Gilliland was cautious in his claims. 'Given the many problems with the data sets,' he said, 'one is not inexorably led to the conclusion that a negative secular solar radius trend has existed since AD 1700, but the preponderance of current evidence indicates that such is likely to be the case.'

With rhythmic variations 11 and 76 years long now identified in the measurements of solar diameter, however, it seems straightforward to interpret the longer trend as part of a similar but longer cycle, posing no real problems for astrophysicists' and geologists' faith in the long-term stability of our nearest star. Gilliland, however, had a parting shot to fire in his *Astrophysical Journal* paper—in 1980, if his analysis was correct, the Sun was approaching a maximum in the 76-year cycle, and began to decline in size once again at about the end of the 1980s. But the accepted textbook value of the solar radius, 959.63 seconds of arc, is based on nineteenth-century observations, which became enshrined in the textbooks at a time when the Sun was close to a minimum of the 76-year cycle. If Gilliland was correct, the true value of the mean solar radius is more like 959.8 seconds of arc. Best of all, though, his claims are testable—existing techniques are accurate enough to measure such changes in the Sun's size directly, and monitoring programmes set up in the wake of these various claims and counter-claims will resolve the issue, one way or another, before the next decade is out.

But while astronomers were still digesting the import of Gilliland's study, and observers were metaphorically girding their loins to meet the challenge of testing these claims, an unobtrusive paper from another theorist added a new twist to the saga of the shrinking Sun. In a paper published in the journal *Astronomy and Astrophysics* just a few weeks after Gilliland's paper appeared in the *Astrophysical Journal*, Carl Rouse of the General Atomic Company in San Diego, California, revived the idea that a decline in the size of the Sun might account for the lack of detectable solar neutrinos. Apparently ignorant of all the fuss about Eddy's claims and the activity they had sparked off—he made no reference to any of the work I have discussed here—Rouse put forward his proposal purely on theoretical grounds. The solar neutrino puzzle can be resolved if the heart of the Sun is 10 per cent cooler than standard astrophysical models imply, and Rouse showed how a slightly different form of mixing of material in some regions of the Sun could produce a cooler core and a contracting outer layer.

So where do we go from here? One path, clearly, leads back to the puzzle of just how the Sun works, and why it is producing so few neutrinos just now. But another path leads inexorably in a different direction, into the detailed study of the workings of the Earth's atmosphere and climatic systems. In his search to determine more about solar variability, Gilliland was forced to unravel the complexities of recent climatic changes on Earth. It is difficult to see how these changes in solar diameter, both the 76-year cycle and the longer-term decline, could fail to have affected the temperature of the globe, but so many other factors also affect the climate that the only hope of detecting the solar 'signal'—and thereby, ultimately, of learning more about what makes the Sun tick—is to subtract out the other main influences from the historical record of changes in temperature in the northern hemisphere.

It might seem a daunting task. But in recent years climatologists have begun an intensive study of temperature changes on Earth, driven in large measure by a concern that the build-up of carbon dioxide in the atmosphere—a by-product of our dependence on energy from fossil fuels—could trap infrared radiation near the surface of the Earth and warm the planet through the greenhouse effect. By and large, climatologists agree that two 'perturbations' have been affecting temperature trends during the present century: the warming influence of this build-up of carbon dioxide, and the variable cooling influence produced when great volcanic eruptions spread dust high into the stratosphere, blocking some of the heat from the Sun. Most of the changes in temperature over the past hundred years can be broadly explained by these two processes at work. But Gilliland found that the real temperature record could be matched much more closely by adding a third factor to the calculations—varying solar heat output tied to the 76-year cycle of solar size variations.

Gilliland was at pains to point out that this does not prove anything about the way external influences affect the workings of the weather machine. His study certainly provides food for thought, though. He took the well-established record of annual changes in the average temperature of the northern hemisphere since 1881, and obtained the best possible fit to this pattern by combining the three external influences, each modified by a scale factor to adjust the agreement between the model and the real world, and with an adjustable time lag in the volcanic and solar influences to allow for the time it takes the temperature of the atmosphere and oceans to respond to outside influences. With so many variable factors, it ought to be possible to provide a fit to almost any curve; the interesting point of the study is whether or not the scale factors you have to put into the equations seem to be telling you anything meaningful about the real world.

The main features that any such model has to explain are a slight warming of the world (at least, of the northern hemisphere, for which good records are available) from the late nineteenth century up to the 1940s, and a subsequent cooling up to the 1970s. Gilliland got a reasonable fit between actual temperature records and his model if he left solar variations out altogether and worked only with the build-up of carbon dioxide and influence of volcanic products on the atmosphere. The early part of the present century was quiet in volcanic terms, and the warming might be explained as a result of dust clearing from the stratosphere. The recent cooling trend coincides with increasing volcanic activity. But in such a variation on the theme, the carbon dioxide influence has to be set very small, only one-tenth of the strength most climate modellers currently accept, otherwise it would have overwhelmed the volcanic influence and caused the Earth to continue to warm up through the 1940s, 1950s and 1960s.

The best fit of all between theory and reality came, however, when Gilliland added the third factor, solar variability. The warming trend was then explained as due to a combination of solar and volcanic influences, with a 24-year lag between the maximum of the solar diameter in 1911 and the peak warmth of the 1930s—the dustbowl era in North America. From about 1940 to the 1970s, in this picture, both the solar and volcanic influences were acting to cool the Earth, more than compensating for the rapid build-up of carbon dioxide, even with the standard greenhouse effect numbers that the climatologists say are the best approximation—that, other things being equal, doubling of the concentration of carbon dioxide in the atmosphere will warm the surface of the earth by 2°C.

All this is very intriguing. It resolves the puzzle of why the Earth cooled even while the concentration of carbon dioxide was continuing to grow exponentially, and all it requires is a peak change in solar luminosity of just 0.28 per cent over the 76-year cycle, producing a maximum influence on surface temperatures on earth of just 0.28°C. These figures are well within the range of possibilities set by observations of the solar constant from the Earth; as Gilliland said, in this picture, 'low temperatures of the last two decades result primarily from a minimum of the solar 76-year cycle.' But while resolving one puzzle about the greenhouse effect, his analysis raises new concern about its future influence on humankind.

If the standard greenhouse effect calculations are indeed correct, as they must be to produce the best possible agreement between Gilliland's model and the real world, then over the next 30 years temperatures are likely to rise by a full degree as ever increasing quantities of fossil fuel are burnt. But now the solar influence is just turning around to contribute a further warming influence up to the year 2010, boosting the greenhouse effect where for the 30 years up to 1990 it was counterbalancing it. The result is a forecast of much more rapid and pronounced warming of the globe than has previously been thought likely, setting in by the end of the 1980s. At first, this might seem beneficial as the 1990s see a return of the excellent conditions for world agriculture that prevailed in the 1950s. Beyond the turn of the century, however, the forecast implies a rapid warming into conditions unseen on Earth for a thousand years or more, heralding a super dustbowl era far worse than the 1930s across the Great Plains of North America.

So, in the space of just three years, the curious puzzle of the shrinking Sun took researchers from studies of dusty old records locked away in the files of the world's most famous observatory to speculations about the internal workings of the Sun and a grim warning of the possible climatic future that will confront the next generation of human beings on an already overcrowded planet. Halley and his successors at the Royal Greenwich Observatory would surely have been fascinated to learn of the unexpected uses to which late-twentieth-century astronomers would put their solar observations. A century or two ago, though, who could have imagined any practical benefit to humankind from such erudite research as studies of the exact path of an eclipse, or painstaking measurements of the angular diameter of the Sun? And how can we imagine the uses which future generations—assuming they survive the coming climatic crisis—might make of the seemingly impractical observations of present-day astronomers—measurements of quasar redshifts, speculations about the black hole in Cygnus, studies of the atmosphere of Jupiter, and all the rest? Perhaps the politicians, the holders of the scientists' pursestrings, should also make a careful study of the curious tale of the shrinking Sun.

seven

The Case of the Missing Neutrinos: 'Curiouser and Curiouser!'

▼

'Is there any other point to which you would wish to draw my attention?'

'To the curious incident of the dog in the night-time.'

'The dog did nothing in the night-time.'

'That was the curious incident,' remarked Sherlock Holmes.

'Some Snarks are Boojums.'

IT TAKES A MIXTURE of Conan Doyle and Lewis Carroll to convey the flavour of the current debate about the puzzle of the 'missing' solar neutrinos. The neutrino saga, as others have pointed out, is at first sight reminiscent of the curious incident of the dog in the night described by Sir Arthur Conan Doyle in his tale 'The adventure of the silver blaze'. The point is that, although solar neutrinos are predicted by nuclear physics theory to be flooding from the Sun in copious quantities, they, like the dog who made no fuss in the night, have yet to make the expected impression on the observer, in this case the world's first solar neutrino detector, a swimming-pool-sized tank filled with cleaning fluid and buried deep underground in a gold mine in South Dakota. The epic story of Professor Ray Davis and his search for solar neutrinos has become part of modern scientific folklore. But it has to be admitted that there is also a touch of Carrollian absurdity about the whole business—if not the hunting of the Snark, then certainly Alice in Wonderland.

The idea of burrowing into the ground to build a detector designed to study the interior of the Sun is curious enough. Even curiouser is the recent revelation that the underground detector may be telling us as much about the origins and fate of the whole universe as it does about the Sun. And then again, it takes a mind like that of the mythical Sherlock Holmes to unravel all the twists of logical argument in the tale and show how what happens in Davis's tank of cleaning fluid really does suggest that the Universe is not destined to expand forever, but may one day collapse back into a fireball reminiscent of the Big Bang of creation.

All good stories have a beginning, a middle and an end. Here and now, 'The case of the missing neutrinos' has only a beginning and a middle: the end is yet to be written. But what we have is intriguing enough.

The beginning, of course, came with the prediction that the nuclear reactions in the Sun ought to be producing vast quantities of fundamental particles known as neutrinos, and with the efforts of Davis and his team to detect them. These efforts have met with partial success—Davis indeed found *some* neutrinos, but not as many as the theories predict. This is, in a way, worse than finding none at all. If the detector detected nothing, the explanation might be that the detector did not work. The fact that some solar neutrinos are detected suggests that the detector does work, but that we don't understand what is going on as well as we thought we did. This has led to the middle of our story—a variety of bizarre speculations offered up to explain the gap between observation and theory. The most outrageous of these was surely the suggestion that the Sun has 'gone out' in the centre, with nuclear fusion reactions temporarily (we may hope!) turned off.

During 1980, however, another possible resolution of the solar neutrino problem appeared, not from astronomical observations or astrophysical theories, but from the particle physicists. Their suggestion, that neutrinos may have mass, solves the solar neutrino problem with one stroke—but it raises equally dramatic and interesting puzzles concerning the nature of the entire Universe. To some theorists, it is a case of 'out of the frying pan and into the fire.' Our understanding of neutrinos and the workings of the Sun may be restored, but only at the cost of a radical revision of our ideas about the evolution of the Universe. To see why, we have to look first at the heart of the problem, the production of neutrinos deep inside the Sun.

To most people, even the idea of the neutrino itself seems a little crazy. The idea of being able to detect neutrinos produced in the heart of the Sun seemed crazier still, until Ray Davis proved it could be done. These singularly elusive particles have zero mass (according to the original theories) and no electric charge, and are identifiable only through a property which the particle physicists dub 'spin'. They are produced in nuclear reactions, but once produced they are extremely reluctant to interact with anything at all. They stream through both empty space and solid matter at the speed of light, unaffected by any conditions much less extreme than those at the heart of a star. And those conditions are extreme indeed. What the astrophysical theories tell us is that stars shine because of nuclear fusion reactions going on inside them. In a star like the Sun, the main process is the fusion of nuclei of hydrogen (protons) into nuclei of helium.

In this way, the Sun, basically a ball of hydrogen 1,392,000 km in diameter, has stayed hot for the 4.5 billion years for which the Earth has, judging by the geological evidence, been in existence. And the models also tell us that the nuclear reactions taking place in the heart of the Sun today operate in a region where the density is 160,000 kg per cubic metre (12 times the density of lead) and the temperature is 15 million degrees.

Trying to catch a neutrino using matter at ordinary densities is much worse than trying to catch the proverbial black cat in a coal cellar at midnight. But it is just because neutrinos are so elusive that detecting them is likely to be highly rewarding. If the theories are correct, any solar neutrinos we could detect on Earth would have been unaffected by *anything* since they were born in the heart of the Sun. Measuring the flow of neutrinos would be like opening a window to the centre of the Sun itself.

Neutrinos offer a chance to check the astrophysical theorists' numbers because the exact temperature and density at the centre of the Sun determine how quickly the fusion reactions which throw off neutrinos take place. Davis's now famous detector was built, in effect, to take the temperature of the heart of the Sun. The detector had to be big by human standards, because it had to contain a lot of atoms to give a chance for a few neutrinos to interact in the tank. And it had to be deep below ground to screen out interference from other particles—cosmic rays. Out of some 400,000 litres of cleaning fluid in the tank, and with no less than 4,000 million suitable neutrinos flooding through each square centimetre of the tank every second (if the theories are correct) the forecast was that Davis should detect no more than 25 neutrinos *each month*! In fact, over more than thirty years of observations, the tank has found no more than eight solar neutrinos each month.

Why has the dog failed to bark? Or, in more technical

terms, why is the Sun producing only one-third of the expected flux of neutrinos? Most of the answers put forward to resolve this puzzle have been astrophysical in nature. In particular, the observations could be brought nicely in line with theory if the temperature in the heart of the Sun is 10 per cent lower than the theorists had calculated. But what the conflict between observation and theory really tells us is that *either* we don't understand astrophysics as well as we thought, *or* we don't understand the particle (neutrino) physics as well as we thought. And the latest turn-up for this particular book has come from the particle physics side.

The whole saga is reminiscent of one of the greatest debates in the history of astronomy, when nuclear physics and astrophysics came into head-on collision some seventy years ago. At that time, the astrophysicists made similar calculations to those they make today, feeding into their equations the known mass and size of the Sun to estimate how hot it must be inside. They came up with the same answer as today, about 15 million degrees. Just as today, they knew—more than half a century ago—that the only way the Sun can have been kept hot throughout the known lifetime of the Solar System is by the efficient and energetic process of nuclear fusion. But nuclear physicists then said that, according to their theories, the hydrogen fusion reaction couldn't work at a temperature of 'only' 15 million degrees and the density at the heart of the Sun.

Drawing confidence from the fact that, whatever the nuclear physicists might say, the Sun exists, the pioneering astrophysicist Arthur Eddington is reported to have told his overconfident nuclear physics colleagues to 'go and find a hotter place'—a polite way of telling them to go to hell! In due course, history proved the astrophysicists right; nuclear physicists found a way to improve their theories to explain how fusion can indeed take place without the need for a 'hotter place'.

The impetus given to nuclear physics by this confrontation with astrophysics was in no small measure responsible for the growth in understanding of both fusion processes and nuclear fission, which eventually led to the atomic bomb, nuclear power and the hydrogen bomb. By highlighting a flaw in nuclear physics theory, the seemingly abstract art of astrophysics had direct and practical implications for the whole of humankind.

It is too early yet to perceive any practical implications of the outcome of the rematch between these two branches of science. It begins, however, to look as if the implications may be profoundly important—as far as abstract knowledge is concerned—and changes in the way we view the world are inevitably followed before long by practical changes, for good or ill.

Ever since it became clear that the Davis neutrino detector was finding about one-third of the expected number of solar neutrinos, a few particle physicists have been offering a neat, but bizarre, way to resolve the conflict between theory and observation. It happens that there are three kinds of neutrino known to occur in nature, one associated with the electron, and called the electron neutrino, one associated with the fundamental particle known as the muon (a kind of 'heavy electron') and one associated with another type of particle called the tauon. The electron, muon and tauon together form a family called the leptons. The neutrinos that are believed to be produced in fusion reactions inside the Sun are all electron neutrinos, and the Davis detector can detect only this kind of neutrino. Just suppose, the theorists suggested, that on the way from the Sun to the Earth electron neutrinos were converted into equal proportions of all three members of the family: one-third electron neutrinos, one-third muon neutrinos and one-third tauon neutrinos. In that case, the Davis tank should detect just about eight electron neutrinos a month—which it does!

The idea is not as crazy as it seems at first sight. In the Alice-in-Wonderland world of particle physics, there are families of particles which behave in this way, decaying into other members of the family through what is called a resonance, with an equal chance that any particular member of the particle family will be produced by the decay. But there is a snag. Neutrinos, according to the original theories, have no mass at all. Massless particles have two important properties. First, they always travel at the speed of light (light 'particles', or photons, are, of course, the classic example of this). Secondly, and much more important for the solar neutrino debate, massless particles do not decay into other members of the same family through this kind of resonance effect.

By saying 'suppose electron neutrinos decay into other kinds of neutrinos', the maverick theorists were saying, in effect, 'suppose neutrinos have mass'. Few people took the idea seriously, even though the requirement is only that the neutrino has some tiny, almost infinitesimal and possibly undetectable mass—anything more than zero. But the possibility nagged away at the particle physicists, increasingly discomfited with every year that went by with no resolution of the solar neutrino problem. By 1980, a few groups around the world were actively seeking to measure either the mass of the neutrino or its decay into other kinds of neutrino. And some of their results upset the applecart by coming out in the affirmative—electron neutrinos *do* have mass and they *do* decay!

It was only in 1956 that neutrinos—long since predicted by the theorists—were detected at all, by Frederick Reines and Clyde Cowan. Appropriately, in 1980 Reines himself, and colleagues from the University of California at Irvine, were among the first to complete experiments that suggest that neutrinos have mass.

The experimental evidence is far from easy to obtain.

First, you need a source of neutrinos—the team used the Dupont Company's 2000-megawatt Savannah River nuclear power plant. This acts as a source of low-energy neutrinos, useless by-products of the nuclear reactions that provide the useful power. Then, you need a detector—and the Davis experiment has shown how hard it is to detect neutrinos. Reines and his colleagues use a pool of heavy water containing 268 kg of deuterium oxide (deuterium is the heavy isotope of hydrogen). Neutrinos can react with protons and neutrons in 'heavy water' in two different ways, producing either one or two free neutrons as a result. Neutrons are easy to detect (compared with neutrinos), and the exact number of times the detectors register a single event, compared with the number of times they record a pair of neutrons, reveals how often the two kinds of reaction take place in the heavy-water bath. The double-neutron event happens only with 'pure' electron neutrinos; the single neutron event happens with any member of the neutrino family. So measuring the ratio of the two kinds of event gives, at last, a measure of how many electron neutrinos change into other varieties in the tank. Reines calculates that in travelling 11.2 metres from the reactor to the detector, roughly half of the neutrinos have changed their spots. Staying in the world of Lewis Carroll, but changing the analogy, it seems that some snarks really are boojums after all!

The fact that neutrinos are observed to change their spots at all is what matters here, since to do so they *must* have mass, however small, and cannot travel at the speed of light, although they can go pretty nearly as fast. The proportion of spot-changers observed doesn't mean a lot, since the experiment involves such a short distance. It would take an astronomical 'path length'—say, from here to the Sun—for the resonance to settle down into an equal distribution of neutrinos among the three members of the lepton family.

As you might expect, the astrophysicists are delighted. Just as in Eddington's day, it seems that they were right all along, and it is the particle physicists who have had to reshape their picture of the Universe to make it fit with astronomical observations. So far, so good. But this is not the end of the story—it isn't even the end of the middle.

The next question, of course, is how much mass each neutrino has. Getting a handle on this is tricky, but a Russian team at the Institute of Theoretical and Experimental Physics in Moscow made a stab at the task. They used a technique which looks at characteristics of the electron emitted by the decay of an atom of superheavy hydrogen—tritium—a process which also involves an electron neutrino. Like the Reines group's experiment, the evidence is indirect and has to be interpreted with care. But it implies that each neutrino has a mass between 12 and 40 electron-volts. The electron-volt (eV) is a very small unit. To put it in perspective, an electron weighs in at 511,000 eV, which is rather less than 10^{-27} (that is, a decimal point followed by 26 zeros and a 1) gram!

Now, this starts to be very interesting on a far greater scale than that of the Sun and Solar System. Each neutrino may be very light. But according to our best theories there must be a very great number of neutrinos in the Universe, since they are produced in profusion by nuclear reactions but are very reluctant to interact with anything after they are produced. In round terms, there ought to be as many neutrinos in the Universe as there are photons of light, about 100 *million* times as many neutrinos as all the protons and neutrons in all the stars and galaxies of the Universe put together. Most of the mass of stars and planets, and therefore most of the mass of galaxies and clusters of galaxies, is in the form of protons and neutrons, forming atomic nuclei. If the mass of a neutrino is just 10 eV, then it is roughly 0.00000001 times the mass of the proton. And in that case all the neutrinos would contain as much mass as all the visible stars in all the galaxies of the Universe put together. If the neutrino's mass is actually more than 10 eV, as the Russian evidence suggests, then *most of the mass of the Universe is in the form of neutrinos.*

It rather looks as if every astronomical observation ever made, except for Davis's solar neutrino detector observations, and every theory constructed from those observations, has been dealing with only a minority of the material in the Universe, the small fraction which makes up visible stars and galaxies. This offers a solution to a problem in astronomy much older than the solar neutrino puzzle.

The problem is the so-called missing mass of clusters of galaxies. By measuring the Doppler redshifts of galaxies in clusters, astronomers are able to estimate the velocities of the galaxies relative to one another and find out whether clusters are flying apart or bound together in a stable fashion by gravity. All too often, calculations based on estimates of the masses of galaxies from their brightnesses imply that clusters are not stable and are indeed flying apart. 'Extra' mass is required in order to explain why clusters exist at all—and a sea of trillions of neutrinos, each contributing its 10 eV or so of mass to the cluster, could be the explanation.

A further implication, however, is far less welcome to most astronomers today. One of the great triumphs of contemporary scientific thought is the Big Bang theory of the Universe. It explains the observation that clusters of galaxies are flying away from one another. It postulates a genuine 'beginning', some 15 thousand million years ago, when the Universe burst into existence in the form we know it, from a superdense state in a singularity, a hot cosmic fireball. Theoretical models for such an expanding Big Bang universe come in many varieties, some of which clearly do not apply to the real Universe, but several of which do. In particular, cosmologists have puzzled over whether there is enough mass in the Universe in the form of galaxies for gravity eventually to halt the expansion and bring about collapse back into another hot fireball reminiscent of the Big Bang of creation. Such a 'closed' Universe is philosophically interesting, because it offers the possibility that the collapse phase could be followed by a 'bounce' and a new Big Bang, followed by a new phase of expansion and collapse, on into the indefinite future and back into the indefinite past. Most cosmologists agree, however, that the best interpretation of studies of clusters of galaxies across the visible Universe is that the Universe is 'open' and will expand forever. In other words, the Big Bang was a unique event. This is rather disturbing on philosophical grounds. But just as massive neutrinos ('massive' simply meaning having mass, even a few electron volts' worth) might provide the glue that holds clusters of galaxies together, so they could provide enough extra gravitational mass to bind the whole Universe together and cause its eventual recollapse. The mass needed is only about 25 eV per neutrino, comfortably in the range estimated by the Russian group.

That might look fine and dandy, especially if you like the idea of a cyclical Universe. But there is a big snag. Early on in the history of the Universe, all those 'massive' neutrinos would have played a very important role in smoothing out the distribution of matter. The theorists ruefully admit that their current best ideas about how stars and galaxies formed simply would not work in a Universe dominated by massive neutrinos. It looks as if a choice may have to be made between massive neutrinos and the favoured theories of the origin and evolution of galaxies in the expanding Universe, and no cosmologist now seems to be persuaded that the experimental evidence for massive neutrinos is yet good enough to tip the balance. Inverting Eddington's old argument, the fact that the Sun, and ourselves, are here at all could be taken as evidence that neutrinos do *not* have mass!

So the saga continues to echo the debate of the 1920s between the astrophysical theorists and the nuclear physicists. The evidence certainly cannot be regarded as conclusive. The end of the story has yet to be written. Both Reines's group and the Russians described their measurements as preliminary, and groups of Swedish and French particle physicists have been unable to find any evidence that neutrinos have mass, or decay in the way the other experiments suggest. Perhaps, after all, the best evidence we have for neutrino resonance is that the Davis neutrino detector finds only one-third of the expected number of solar electron neutrinos each month. But there is at least a hint here that, compared with seventy-years ago, the boot may be on the other foot. Last time, the nuclear physicists had to change their cherished theories in order to accommodate observed astrophysical reality; this time around—just possibly—the astrophysicists are going to have to change some of their cherished views, on how galaxies behave and where stars come from, in order to accommodate the observed particle physics reality.

It's a long way from a detector buried in a goldmine in South Dakota to the heart of the Sun, and even farther to the outer limits of the visible Universe. Yet it looks at least possible that Ray Davis's tank of cleaning fluid may settle one of the deepest and most puzzling issues of science: the ultimate fate of the Universe. The incident of the dog in the night may have been curious, but to describe the ins and outs of this particular astronomical saga, once again only the world of Lewis Carroll seems appropriate—'Curiouser and curiouser!', as Alice cried, early on in her adventure down that other, even more famous hole in the ground. We haven't yet seen the end of the solar neutrino saga, but we may at last have seen the end of the beginning of the story.

eight

Stardust Memories

▼

SUPERNOVAE ARE THE GREATEST of all stellar explosions, events so powerful that for a brief period a single star will emit, in its death throes, as much light as all the stars of the Milky Way put together. Such events are rare. Our Sun is not fated to become a supernova, but it was born out of the debris of supernova explosions of the distant past, when our Milky Way Galaxy was young. Apart from hydrogen, every atom in your body, and every atom on Earth except for hydrogen and helium (there is no helium in your body) was manufactured inside stars and then expelled into space by supernova explosions. They laced the clouds of hydrogen and helium from which the Sun and its family of planets formed.

Over three decades, beginning in the 1950s, theorists had developed what seemed to be a satisfactory understanding of supernova explosions, based on their understanding of the laws of physics, on observations of such explosions in distant galaxies and of the debris from old supernova explosions in our own Galaxy, and on computer models of how stars work. But until 1987 they had no means of checking this understanding directly. The explosion of a star known as Sanduleak –69° 202 to become a supernova first visible from Earth on the night of 23/24 February 1987 was, therefore, possibly the single most important event in astronomy since the invention of the telescope. The event, dubbed SN1987A (denoting the first supernova observed in 1987), took place in the Large Magellanic Cloud, a galaxy close to our own Milky Way and part of the system of galaxies, held together by gravity, known as the Local Group. At a distance of 180,000 light years (just next door, by cosmological standards), SN1987A was by far the closest supernova to have occurred since 1604, when the last known supernova in our own Galaxy exploded, just before the development of the astronomical telescope. It was close enough to be studied in detail by a battery of instruments, including conventional telescopes on mountaintops, X-ray detectors on board satellites in space and neutrino detectors buried deep beneath the ground. Both in broad outline and in most details, those observations showed, over the years following the outburst, that the astronomers did have a good understanding of how supernovae work. Although some details did not match up to expectations, there were no major surprises. It seems that we do, indeed, understand what happens in a supernova.

The discovery of SN1987A caused huge excitement among astronomers, spilling over into the general press and making the cover of *Time* magazine on 23 March that year. The reason for all the excitement was partly the importance of supernovae themselves—they are the biggest explosions that have taken place since the Big Bang in which the Universe was born, and the source of all the heavy elements—and partly their rarity. Only four definite supernova explosions have been observed in our Galaxy over the past thousand years, and the last one visible to the unaided eye (and that only just) blew up in the Andromeda Galaxy, two million light years away (ten times farther than SN1987A) and was visible as long ago as 1885.

The last supernova seen in the Milky Way occurred in 1604. Curiously, though, modern astronomers have found the remains of a supernova that ought to have been visible from Earth in the middle of the seventeenth century, but which nobody seems to have noticed at the time. It is called Cassiopeia A. Although the 1604 supernova was studied in detail by Johannes Kepler, frustratingly for astronomers today his records were entirely based on observations with the unaided eye. The supernova was visible from Earth just five years before Galileo first applied the telescope to the study of the heavens. Before the astronomical telescope was invented, supernovae visible from Earth had been popping off in our Galaxy at a rate of about four every thousand years. By blind chance, two had been visible in the span of a single human lifetime, in 1572 and 1604. But in all the time from 1604 to 1987, as telescopes lay in wait for their prey, the only supernova that could (just) have been seen by the unaided eye was the one spotted in the Andromeda Galaxy at the end of the nineteenth century. That's why SN1987A caused so much excitement among astronomers. It wasn't quite in the Milky Way, but in the galaxy next door, and it was visible to the naked eye. It could be studied in unprecedented detail by all the instruments that now exist to supplement Galileo's simple telescope.

The supernova was first seen by a young Canadian astronomer, Ian Shelton, from Las Campanas Observatory in Chile, but what turned out to be key observations had been made even before the supernova was noticed, by astronomers taking routine photographs of the Large Magellanic Cloud. Robert McNaught, in Australia, photographed the brightening star about 16 hours before it was identified as a supernova, using a large astronomical camera known as a

Schmidt telescope—but the photographs were only developed and studied after the news from Chile reached Australia. About three and a half hours later, two astronomers testing a new piece of guiding equipment on a telescope in New Zealand just happened to pick the Large Magellanic Cloud as the target for their test photographs. Together with the observations from Chile the night before the supernova burst into view, these photographs helped to establish the timing of the event, and the speed with which the progenitor star, Sanduleak -69° 202, flared up. Even better for astronomers, this is the first time a star that became a supernova has been identified on old photographic plates, so that they know in some detail what it was and what it was doing before it flared.

All of this helped astronomers to test their theories of how supernovae work. The key theoretical insight actually dates back to 1934. At that time, less than two years after the discovery of the neutron, Walter Baade and Fritz Zwicky made the dramatic suggestion that 'a supernova represents the transition of an ordinary star into a neutron star'. But although half a century of observations of distant supernovae and theorizing had filled in the details of how that might happen, the theories could only be tested fully by studying a nearby supernova at work.

By the late 1980s, astronomers were satisfied, from their studies of supernovae in other galaxies, that there are two basic, different types of supernova. In each case, an ordinary star is indeed converted into a neutron star, releasing gravitational energy as it shrinks. The more it shrinks, the more energy is released. The total energy released is enormous because a neutron star is so small: it contains as much mass as our Sun, but packed into a sphere no more than a few kilometres across, a volume comparable to that of a mountain on Earth. Such a star will form from any lump of matter that is no longer kept hot by nuclear fusion in its heart (a dead star) provided its mass is a little more than a critical amount (actually slightly bigger than the mass of our Sun). This occurs when the inward tug of gravity overwhelms the forces that give atoms their structure. If the mass is *much* bigger than this, even neutrons are crushed out of existence by gravity, turning the dead star into a black hole. The range of masses for stable neutron stars is, therefore, only from a little over the mass of our Sun to about 2 or 3 solar masses.

The first way to make a supernova (type I) involves a cold, dead star which has less than the critical amount of mass. It then gains additional matter from a nearby companion. Such a star starts out as a white dwarf, a dead star with about the mass of the Sun, maybe a little less, contained in a volume the size of the Earth. It is the fate of our own Sun to end its life as a white dwarf, because it does not have enough mass to become a neutron star and it has no companion from which to steal mass. A star like the Sun which has become a white dwarf and orbits around another star can gain mass, however, by tugging streamers of gas off its companion through tidal forces and swallowing the gaseous ribbons. When its mass reaches the critical value, the atoms of which the star is made will collapse, electrons being forced to merge with protons to become neutrons. The star, now more massive than the Sun, will shrink from the size of the Earth to the size of a mountain and will release the appropriate amount of gravitational energy in the process.

But that is not what happened in SN 1987A. There is another way to make a supernova, known as type II. This happens, according to theory, when a very massive star near the end of its life runs out of nuclear fuel to keep its heart hot. A star like the Sun keeps hot by 'burning' hydrogen to make helium, in a process known as nuclear fusion—the same process that operates in a hydrogen bomb. Later in its life, helium itself is burned to make carbon, and so on. But when it has no more nuclear fuel to burn, the inner part of such a star, already with more than the critical mass needed to make a neutron star, collapses all the way to the neutron star state, without stopping off as a white dwarf. Comparably greater amounts of energy are liberated—at least a hundred times as much energy, in a few seconds, as the Sun has radiated in its entire lifetime—blasting the outer layers of the star outwards at speeds of around 20,000 km per second (actually 17,000 kilometres km per second in the case of SN 1987A) and triggering a wave of nuclear reactions which manufacture heavy elements that can be formed naturally in no other way.

SN 1987A was a type II supernova, the most energetic kind of stellar event that can ever take place. And, because the original star that exploded has been identified, astronomers can reconstruct the history of the supernova from the time that star was born right up to the dramatic events observed in 1987.

That story is reconstructed, of course, with the aid of the computers that model how stars work. Different researchers have developed slightly different models. They tell slightly different stories, although the broad outlines are always the same. The outline I give here is based on the models used by Stan Woosley and his colleagues. Woosley, a supernova expert, works at the University of California, Santa Cruz. According to his model, the star in which we are interested was born only about 11 million years ago, in a region of the Large Magellanic Cloud particularly rich in gas and dust. Because the star contained about 18 times as much matter as the Sun, it had to burn its nuclear fuel more quickly in order to provide enough heat to hold itself up against the inward tug of gravity. So its fuel was exhausted more quickly than the fuel of a star of 1 solar mass, and it blazed about 40,000 times brighter than the Sun. In just 10 million years it had burned all the hydrogen in its core into helium—essentially the same process that keeps the Sun hot today. As a result, the core slowly shrank and got hotter until the next set of nuclear reactions, which convert helium into carbon and release energy, could begin.

During this phase of its life, such a massive star becomes a supergiant, the outer layers swelling up to stretch across a distance roughly the same size as the diameter of the Earth's orbit around the Sun. One of the surprises that astronomers found when they examined old photographs of SN 1987A's progenitor, Sanduleak -69° 202, was that the star was actually not a red supergiant but a blue supergiant, a smaller and hotter type of star. The outer parts of the star had contracted again slightly, perhaps as recently as 40,000 years before the explosion. This does not affect the basic understanding of type II supernovae, but it gives the theorists plenty of interesting detail to get their teeth into. A favoured explanation at present is that this late shrinking of the outer part of the star has to do with the fact that the Large Magellanic Cloud, unlike our own Milky Way Galaxy, contains only relatively modest amounts of elements heavier than helium. One of those elements deficient in stars of the Large Magellanic Cloud, oxygen, helps to make a red supergiant swell up, because a little oxygen in the outer part of the star absorbs radiation that is trying to escape, holding it in and making the star swell like an inflating balloon. With less oxygen present, once such a star reaches the stage of its evolution where the outward flow of radiation drops slightly, the 'balloon' might deflate again. While helium burning was going on, the star was probably a red supergiant, but helium burning could sustain the star only for about a further million years after hydrogen burning in the core ended.

In the last few thousand years of its life, Sanduleak –69° 202 must have gone through its remaining possibilities for energy production with increasing speed. Carbon, itself a product of helium burning, was converted into a mixture of neon, magnesium and sodium. Neon and oxygen (another product of helium burning) 'burned' in their turn, and at the end, nuclear fusion reactions were consuming silicon and sulphur in the heart of the star, while all the other nuclear fuels were being burned in successively cooler layers working outwards from the centre. All the while, the pace of change quickened. According to the calculations made by Woosley and his colleagues, helium burning lasted nearly a million years, carbon burning only 12,000 years. Neon kept the star hot for 12 years. Oxygen provided the necessary energy for a mere four years, and silicon was burnt out in a week. And then, things began to get really interesting.

Silicon burning is the end of the line even for a massive star, because it produces a mixture of nuclei, including cobalt, iron and nickel, that are among the most stable arrangements it is possible for protons and neutrons to form. Sticking lighter nuclei together to make iron nuclei releases energy. But sticking iron nuclei and lighter nuclei together to make heavier elements uses up energy. Indeed, heavier elements may fission, splitting to form nuclei more like those of iron and giving up energy in the process. There is a kind of natural energy valley for nuclei, with iron at the bottom and light elements up one side of the valley while heavier elements up on the other slope of the valley. All nuclei would 'like' to roll down the valley and become iron, light ones through the fusion route and heavy ones through the fission route. In this sense, iron and nickel are the most stable nuclei. So, where do elements heavier than iron (lead, uranium, and all the rest) originate? In supernovae, like SN 1987A. And although that statement was well founded in scientific calculation before February 1987, it has only been proved correct by studies of SN 1987A itself.

Particle theorists, drawing on the studies by their experimental colleagues, can explain how elements heavier than iron can be produced, provided the nuclei are bathed in a sea of neutrons. And neutrons are one thing that a supernova produces in abundance.

Most of the elements more massive than iron, as well as some of the isotopes of less massive elements, are produced when nuclei built up by nuclear fusion processes capture neutrons from their surroundings inside a star. Any free neutron is itself unstable, and emits an electron by beta decay, turning into a proton, if left to its own devices for a few minutes. So the neutrons involved in these capture processes have to be freshly released by other nuclear interactions. This is no problem inside a star where nuclear burning is going on. For example, every time one nucleus of deuterium and one of tritium fuse to produce helium-4, one neutron is released. This and similar reactions inside stars provide a profusion of neutrons—as many as a hundred million in every cubic centimetre of the interesting region of a star—which may interact with other nuclei.

Adding a single neutron to a nucleus increases its mass by one unit, but does not change its electric charge or its chemical properties—it becomes a different isotope of the same element. In many cases, however, the newly formed isotope is unstable, and given time (in some cases, a few seconds; in others, a few years) it will eject an electron by beta decay, converting one of its neutrons into a proton and becoming a different element. The whole process may repeat when the same nucleus captures another neutron. This step-by-step build-up of heavy elements, in which a nucleus has time to convert into a stable form in between interactions with neutrons, is known as the slow, or s-process of neutron capture.

But when large numbers of neutrons are available, which certainly happens as a result of explosive interactions occurring during the early stages of a supernova, there may be so many neutrons around that a nucleus can capture several of them before it has time to spit out an electron, or decay in some other way. A density of a mere hundred million neutrons per cubic centimetre is nowhere near enough to make this happen; it requires a density of about three hundred billion billion (3×10^{20}) neutrons in every cubic centimetre of star stuff. The result, when these enormous neutron densities are briefly achieved as a supernova explodes, is a rapid build-up of elements and isotopes which have a surplus of neutrons and which are almost all unstable. This is the rapid, or r, process of neutron capture. Once the wave of neutrons has been absorbed, the unstable, neutron-rich nuclei that are left behind will decay into stable nuclei, losing neutrons (converting them into protons) and becoming more like the isotopes produced by the s-process. Many isotopes are produced by both processes. A handful of stable, slightly neutron-rich nuclei are produced only by the r-process and subsequent beta decays. And just 28 isotopes, astrophysicists calculate, can be produced only by the s-process.

In a diagram of the elements which plots the number of neutrons in a nucleus against the number of protons, stable isotopes lie on a roughly diagonal band along which the number of neutrons is slightly greater than the number of protons. Elements formed by the s-process (and by the ultimate beta decay of r-process elements) lie on a zig-zag track through this 'valley of stability'; unstable isotopes produced by the r-process lie far off to the right, in the neutron-rich half of the diagram, and as they decay they shift towards the bottom of the valley of stability, 'raining down' on the s-process elements. Both processes end for very massive elements where nuclei are split apart either by alpha decay (emitting a helium nucleus) or by fission (producing two roughly equal nuclei each with about half the mass of the one that splits).

The studies of SN 1987A showed just how well astrophysicists really do understand the way elements are built up in supernovae.

Except in the case of our Sun, where neutrino studies may be providing a direct clue to conditions in the deep interior, we cannot study any of the nuclear processes at work inside stars directly. The observations that provide both the input to theories of stellar astrophysics and the tests of those theories are indirect studies of material expelled from inside stars. First, the material has to be processed inside a star, carried to the surface and ejected into space. Then it can be studied from Earth by the way in which the elements produced either radiate or absorb light. Everything has to fit together—and it does.

Clouds of gas known as planetary nebulae (a misnomer, as they have nothing to do with planets) produced by relatively small stellar explosions—mere novae—also show the 'right' mix of elements and isotopes. But as well as providing the biggest input of energy to make heavy elements, supernovae also provide the biggest blast to blow material out into space, giving astronomers their best opportunities to study star stuff. Many old supernova remnants have, indeed, been identified and studied by spectroscopy. But there's a snag. A cloud of gas blown out into space by a supernova explosion sweeps up gas and dust from between the stars as it moves through space. So when astronomers study the glowing cloud of material today, hundreds or thousands of years after the supernova explosion lit up the night sky for our ancestors, they cannot untangle the information they want about the elements produced in the supernova itself. SN 1987A, caught on photographic plates before, during and after its explosion, was a different story.

According to astrophysical theory, just before the supernova exploded all the standard nuclear reactions leading up to the production of iron-group elements were going on in shells around the core, and in addition, the s-process should, theorists were confident, have been at work in the region of the star rich in carbon and oxygen. Silicon burning, remember, had held the star up against the inward tug of gravity for just about one week, and had left a core composed of the most stable nuclei—iron, nickel and the rest—incapable of releasing energy *either* by fusion *or* by fission (although, as we shall see, some of those iron-group nuclei can decay to iron itself). After 11 million years, the heart of the star was left with no means of support, and it collapsed, in a few tenths of a second, into a lump no more than 100 km across. During this initial collapse, very energetic photons ripped iron nuclei apart, undoing the work of 11 million years of nuclear fusion processes, and electrons were squeezed into nuclei by such enormous pressures that beta decay went into reverse, converting protons into neutrons. Gravity provided the energy for all this. All that was left was a ball of neutron material, essentially a giant 'atomic nucleus', about 200 kilometres across and containing nearly one and a half times the mass of our Sun.

The squeeze of the infalling material was so great that at this point the centre of the neutron ball was compressed to densities even greater than those of the nucleus of an atom. Then, like a golf ball being squeezed in an iron grip and then released, it rebounded, sending a shock wave out through the ball of neutron stuff and into the star beyond. Material from farther out in the core of the star, plunging inwards at a speed roughly one-quarter of the speed of light, met the rebounding shock from the core of neutron stuff and was literally turned inside out, becoming a shock wave racing outwards through the star. It was this shock wave that blew the star apart—but not before a flood of neutrons emitted by all this activity had caused a sizeable production of heavy elements through the r-process. The shock moves outward at about 2 per cent of the speed of light, taking a couple of hours to push the outer layers of the star into space and light up the star visibly.

While all this was going on, even though the iron core of the star had been converted into a ball of neutrons, according to theory there should have been a massive burst of nuclear reactions farther out in the star, in the hot, high-pressure shock wave, producing iron-group elements. Most of the elements produced inside the star by such fusion reactions are made, in effect, from successive additions of alpha particles (helium-4 nuclei, each made of two protons and two neutrons combined together), and have equal numbers of protons and neutrons in their nuclei. Carbon-12 (six protons, six neutrons) and oxygen-16 (eight protons, eight neutrons) are typical examples. When these nuclei are processed by explosive interactions, according to theory most of the material is converted into nickel-56, which has 28 neutrons and 28 protons in each nucleus. But nickel-56 is unstable: it decays radioactively, emitting positrons as protons which are converted into neutrons (inverse beta decay). The first step in this decay has a half-life of just over six days, and produces cobalt-56; the cobalt-56 then decays into iron-56 (26 protons, 30 neutrons) with a half-life of 77 days.

The unstable nickel-56 has been built up by the input of gravitational energy from the collapse of the core of the supernova. When it decays, it gives up some of that borrowed energy. The standard theory of supernovae, developed before SN 1987A was seen to explode, predicted that almost all of the energy radiated by the star during the first hundred days of its life as a supernova would come from the decay of cobalt-56 into iron-56. This decay follows a characteristic pattern, a decreasing exponential curve. The fading of the supernova itself followed exactly the predicted curve. During that first hundred days, 93 per cent of the output of energy from the supernova was indeed provided by the decay of cobalt-56. The slow fade of the supernova was still following the appropriate curve in 1990, three years after Shelton first noticed the ignition of the supernova. Astrophysicist Roger Tayler, of the University of Sussex, believed that these observations of cobalt decay were 'probably the most important and exciting ones concerned with the origin of the elements, confirming that the theoretical model is broadly correct'.

It wasn't just the 'light curve', as it is called, to which he was referring. As the material expelled by the supernova moves out into space, successive layers of its interior are revealed to the telescopes of patiently watching observers, in a kind of cosmic striptease. Eventually, they could see material coming out from the regions where the explosive nuclear interactions should have taken place—and what their spectroscopic studies revealed was characteristic lines associated with nickel-56, just as expected, indicating (after allowing for the decay that had already taken place by the time this part of the star could be observed) that as much nickel-56 as the equivalent of 8 per cent of the mass of the Sun had been manufactured in the supernova—closely in line with theoretical calculations. The spectroscopic studies also reveal the presence of barium, strontium and scandium—all s-process elements produced before the star became a supernova. Studies of helium and nitrogen in the outermost layers of the expanding cloud of material around the supernova are helping astrophysicists to improve their understanding of how material produced by more routine nuclear fusion when the star was young gets mixed up to the surface.

Of course, there were also surprises. Details of the behaviour of SN 1987A do not in every case match precisely with the details of the theories, and there is ample room for astronomers to refine their understanding of how stars like this explode. But the mention of new insights into the way carbon and associated elements are produced and mixed into the Universe provides a cue to bring my present discussion to an end. These, after all, are the elements of which we are, in large part, made—carbon, oxygen and nitrogen have a key importance for life as we know it. And observations of the spectra of these elements in the expanding cloud of material around SN 1987A provide a reminder that while such an explosion marks the death of a star, it is quite literally the beginning of the story of life forms like ourselves. We would not be here, wondering about such puzzles, if it were not for those previous generations of supernova explosions that scattered their share of carbon, nitrogen, oxygen and other elements through interstellar space billions of years ago. Everything on Earth is a stardust memory, made from the scattered remains of dead stars. As far as life forms like us are concerned, in fact, my story ends—in the beginning.

nine

Puzzling Pulsars

▼

ON 24 FEBRUARY 1968, *Nature* carried the announcement of the discovery of pulsars. Then, just three of these objects were known, and their nature was a puzzle. Today more than 550 are known, and they are among the most important phenomena in astrophysics.

Although pulsars were first identified by chance in 1967, that discovery had its roots in a scientific development carried out during the Second World War by scientists who had been diverted from more abstract research. That development was radar. Before the war, astronomers only had observations of the Universe made at visible wavelengths, using optical telescopes. Although the fact that radio waves from space could be detected on Earth had been noticed in the 1930s (by Karl Jansky, working at the Bell Laboratories in New Jersey) there was no time for radio astronomy to develop properly before the war broke out. During the war, radar systems along the coast of the English Channel suffered from interference which was identified as radio noise coming from the Sun, and this fanned the interest of scientists involved in radar work. After the war, in many cases initially using war surplus radar equipment, some of them began to probe the Universe at wavelengths longer than those of visible light, in the radio part of the electromagnetic spectrum. This new window on the Universe transformed astronomy in the 1950s.

Radio astronomy has one great advantage over optical astronomy. The bright blue light of the sky, which makes the stars invisible by day, is actually blue light from the Sun that has been bounced around the Earth's atmosphere ('scattered') by tiny particles in the air, so that it comes at us from all directions. Red light, with longer wavelengths, is not scattered anywhere near so much, and that is why sunsets are red. This kind of scattering does not happen at radio wavelengths. Radio telescopes are not dazzled in the way that our eyes, or photographic equipment attached to telescopes, are dazzled during the day, provided you don't point the radio telescope right at the Sun. In any case, the Sun is nowhere near as bright in radio light as it is in visible light. So radio astronomers can observe interesting objects in the heavens 24 hours a day and don't have to shut down when the Sun is above the horizon.

In fact, the Sun does influence radio waves coming to us from space. But astronomers are cunning enough to make use of this interference with the signals they receive to find out more about the objects in space that emit the radio waves. There is a constant stream of material escaping from the surface of the Sun and blowing out into space and across the Solar System. This is a very tenuous cloud of gas known as the solar wind. The atoms in this wind are not electrically neutral; even at the surface of the Sun, conditions are energetic enough to remove electrons from the outside of the atoms. The solar wind is therefore an electrically charged plasma, but it is much more tenuous than the hot plasma that exists inside a star like the Sun. The density of this solar wind plasma varies—the clouds of material moving out from the Sun are not uniform. These different clouds make radio waves passing through the plasma vary slightly in strength—they 'twinkle', or scintillate, in just the way that variations in the atmosphere of the Earth make starlight twinkle.

Stars are affected in this way only because their images are very small—just points of light. Planets, which appear as tiny disks in the sky, do not twinkle. The tiny fluctuations are averaged out over the visible disk of a planet. Of course, stars are really bigger than planets. They only look like points of light, instead of disks, because they are so far away. The same rule applies to radio sources affected by the solar wind, but it provides extra information about radio sources because, unlike stars, some of them are so large that they do show up as extended features on the sky, not just as points. In the early days of radio astronomy, it was difficult to get a precise 'picture' of a radio source, a detailed map equivalent to a photograph of a star. This is less true today, but then it was not always obvious whether the noise was coming from a point source or an extended source. Sources that twinkle, however, are definitely point sources; those that don't are extended objects. Twinkling radio sources, we infer, must be a very long way away.

It works both ways. The fact that distant radio sources twinkle also reveals information about the nature of the solar wind, and it was this line of attack that led a young radio astronomer, Antony Hewish, to begin investigating such scintillating radio sources at the new radio astronomy observatory in Cambridge in the 1950s. Although he began by using scintillations to probe the solar wind in the 1950s, he moved on to using scintillations to investigate the nature of radio sources, using a government grant of just £317,000 to build a new radio telescope. The pioneering radio astronomer Sir Bernard Lovell has described this award of funds as 'one of the most cost-effective in scientific history'. With this new telescope one of Hewish's research students, Jocelyn Bell, discovered the first pulsar in 1967.

Bell (now Dr Jocelyn Bell Burnell) was born in Belfast in 1943, and she graduated from the University of Glasgow in 1965. During the next two years, she started her Ph.D. studies in Cambridge and worked on the construction of Hewish's new telescope. It scarcely resembled the kind of bowl-shaped antennae that the term 'radio telescope' immediately conjures up in the minds of most people. You need a special kind of telescope to observe the scintillation of radio sources, for it has to be able to respond to very rapid fluctuations in the strength of the radio noise coming from space. Your eyes, for example, can see stars twinkling because they react very quickly to changes in starlight—in 'real time', to use the computer jargon. A photographic plate, on the other hand, exposed for several minutes (or several hours) builds an image over all of that time ('integrated' for all that time). The photograph reveals stars fainter than you can ever see with your unaided eyes, but it will never show twinkling. In the same way, a radio telescope that integrates the signal from a distant object for a long time might be useful in locating the object, but it will never reveal scintillation. The new scintillation telescope developed by Hewish was designed to operate in real time, with a very rapid response to fluctuating signals.

It was more like an orchard than the everyday image of a telescope. A field covering 18,000 square metres was filled with an array of 2,048 regularly spaced dipole antennae. Each dipole (a long rod aerial) was mounted horizontally on an upright, so that it was a couple of metres above the ground, making a letter 'T' with a wide crossbar. The length of the crossbar was chosen to fit the wavelength of radio noise that Hewish was interested in observing. In fact, the crossbar was slightly below the top of its supporting pole. Mixing the analogy, each of the dipoles, mounted across its support, looked like the crossed yard of a square-rigged sailing ship, slung across its mast. All these antennae had to be wired up correctly so that any radio noise they picked up would be combined into one signal. That was fed into a receiver where the fluctuating signals were recorded automatically in pen and ink as wiggly lines on a long strip of paper continuously unrolling from a chart recorder. By varying the way in which the inputs from each of the 2,048 antennae were added together, this system made it possible to sweep a strip of the sky running north and south, and directly overhead at Cambridge. In order to do this, the wiring had to be just right. This tedious wiring task was obviously just the job for a research student.

The aim of the project was to identify very distant radio sources, known as quasars, by their scintillation. By the summer of 1967, the new telescope was up and running, and it detected scintillating radio sources, as intended. You can't 'steer' a field full of antennae the way you can move a dish antenna around to look at different parts of the sky, but with the system Hewish had in place, which Bell was using for her real doctoral work, you can let the rotation of the Earth sweep everything around so that you cover the whole sky once every 24 hours. Because the scintillation is caused by the solar wind, it is strongest when the Sun is high in the sky. The Cambridge team left its system switched on permanently, however. Once built, the instrument cost very little to run, and you never know when you might find something interesting and unexpected.

On 6 August 1967, that is exactly what happened. Each sweep around the sky produced a strip of chart 30 metres long, adorned with three wiggly lines from the pen recorders. As the telescope swept around the sky, any particular source would be 'visible' to it for just three or four minutes, at a time when it was directly overhead. It was Bell's job to examine kilometres of chart to find anything that looked interesting in the wiggles. When she studied the chart for 6 August, she found a tiny fluctuation, about one centimetre long, corresponding to a faint source of radio noise observed by the telescope in the middle of the night, when it was pointing in the opposite direction from the Sun. It couldn't be scintillation. Most probably it was interference from some human activity. Bell marked what she called the 'bit of scruff' on the chart, and ignored it.

But the scruff kept coming back—almost, but not quite, at the same time every night. In September, Bell had enough information to show that the scruff was always coming from the same part of the sky, reappearing at intervals not 24 hours apart, but 23 hours and 56 minutes. This was an important clue. The Earth's motion in its orbit around the Sun shifts the apparent passage of the stars overhead so that an appearance repeats every 23 hours and 56 minutes, not every 24 hours. Just when Bell and Hewish had decided that they had found something interesting, and set up a high-speed recorder to monitor the fluctuations of the scruff, it faded from view for a few weeks. But in November it was back, and the new recorder showed that the scruff was actually a radio source fluctuating regularly with a period of 1.3 seconds.

This was such a surprise that Hewish dismissed it, once again, as interference from a human source of radio noise, despite the fact that it stayed in the same place among the fixed stars. Nobody had ever seen any astronomical object vary that rapidly. The most rapidly varying stars known in 1967 fluctuated with periods of about eight hours. But continuing observations gradually ruled out any possibility of human interference and showed that the pulses themselves were extraordinarily precise, recurring every 1.33730113 seconds exactly. Each lasted for just 0.016 of a second.

Together, these measurements showed that the source of

the pulses must be very small. Because light travels at a finite speed, and nothing can travel faster, fluctuations in any signals from any source can keep in step with one another only if the source is small. That way a light ray can travel right across it during the interval between pulses.

It works like this. If a star like the Sun is so far away that we can only see it as a point of light, the brightness of the star, as we see it, depends on the brightness of different patches of the surface of the star added together. You can imagine that the northern hemisphere of the star might get 10 per cent brighter while the southern hemisphere gets 10 per cent dimmer. The result? We would see no change in the total brightness of the star. We would only see the brightness fluctuate if the whole star got dimmer and brightened in step, and that can happen only if the variations happen slowly enough. There must be enough time for some sort of message to get from the north pole to the south pole. In effect, it telegraphs, 'I'm about to start getting brighter, and you had better do so as well.' The 'message' might be a regular variation in pressure, or a repeated change in the way convection is carrying energy outward from inside the star. Here's the point. Whatever the physical cause of the variation, its influence can spread only at the speed of light, or less. The whole star can respond in step to a disturbance only if it is small enough for the appropriate message to reach every part of it before the message changes. Otherwise, some parts will be getting brighter and others dimmer, in a confused mess of variations. A precise pulse 0.016 seconds long, repeating precisely every 1.33730113 seconds, can only come from something very small indeed—something about the size of a planet, or even less.

Hewish and his team had to face the very real possibility, as of November, 1967, that what they had detected was indeed a signal coming from a planet—a beacon radiated by another intelligent civilization. Tongues only slightly in their cheeks, they speculated among themselves that they might have made contact with little green men, and dubbed the source LGM 1. Hewish decided to keep the lid on news of the discovery until they had carried out more observations. It was just as well that he did.

All the astronomers in Cambridge knew that the radio people at the Cavendish were up to something, but just what it was, nobody could prise out of them. Well, they thought, no doubt the Hewish team would tell everyone in its own good time.

Just before Christmas, Bell found another piece of scruff, coming from another part of the sky. This one turned out to be a similar source, pulsing with comparable precision to LGM 1, but with a period of 1.27379 seconds. Soon there were two more to add to the list, with periods of 1.1880 seconds and 0.253071 seconds, respectively. The more sources were discovered, the less likely the little green man explanation seemed. In any case, careful observations of the first of these objects had shown, by the beginning of 1968, no trace of the variations that you would expect if they were actually coming from a planet in orbit around a star. They must, after all, be natural. The LGM tag was quietly dropped, and Hewish decided it would be safe to go public—first with a seminar in Cambridge, to let the rest of the astronomers there in on the act, and then, almost immediately, with a paper in *Nature* (the issue dated 24 February 1968), announcing the discovery to the world.

The radio astronomers had indeed discovered a new kind of rapidly varying radio source. The title of the discovery paper was 'Observation of a rapidly pulsating radio source', and the term 'pulsating radio source' soon gave rise to the name 'pulsar', which stuck. But what *were* these pulsars that Bell had discovered?

With the announcement of the discovery of pulsars, all hell broke loose among the theorists. A whole new kind of previously unsuspected astronomical object had been discovered, and somebody was going to make a name by finding an explanation for the phenomenon. In the discovery paper, Hewish, Bell and their colleagues pointed towards what seemed the obvious possibilities. If the radio pulses were being produced by a natural process, not by an alien civilization, they had to be coming from a compact star: nothing else could supply the energy required to power the pulses. A star the size of a planet like the Earth had to be a white dwarf. Anything smaller (also allowed by the rapid pulsations) would have to be a neutron star.

At that time, astronomers knew that white dwarfs existed. But neutron stars—objects so dense that the mass of the Sun would be packed into a sphere a mere 10 km across—were regarded as a wild theoretical speculation.

Many stars were known to oscillate, or vibrate, breathing in and out as a result of regular variations in the processes producing energy inside them, and varying in brightness as a result. Maybe this could also happen in compact radio stars. 'The extreme rapidity of the pulses,' wrote the Cambridge team in their *Nature* paper, 'suggests an origin in terms of the pulsation of an entire star.' And they pointed out that the rapid speed of the fluctuation meant that the star doing the pulsating had to be either a white dwarf or a neutron star. There was one snag. Although calculations of the pulsation periods of white dwarfs had been carried out by theorists in 1966, the basic periods they came up with were no less than 8 seconds, a little too long to explain the pulsars. On the other hand, even a simple calculation showed that neutron stars would vibrate with periods much shorter than those of the first pulsars discovered, around a few thousandths of a second. White dwarfs looked the better bet, if some way could be found to allow them to vibrate a little more rapidly than the earlier calculations had suggested.

Further calculations showed that by allowing for the effects of rotation, white dwarfs might vibrate as rapidly as ten times a second. But as more observations of more pulsars were made by radio astronomers around the world (a couple of dozen by the end of 1968; scores more by now), it became clear that pulsars could not possibly be white dwarfs after all.

The fastest possible vibration period, using unrealistic amounts of rotation, was still greater than the periods of some of the new pulsars being discovered. One discovery was particularly significant. It was made by astronomers using the 100-metre dish antenna at Green Bank, West Virginia—just about any kind of radio telescope can observe pulsars, once you know what to look for. They found a pulsar flicking on and off thirty times a second, near the centre of a glowing cloud of gas known as the Crab Nebula.

The high speed of the Crab Pulsar, as it became known, was already enough to put the white dwarf model in trouble (and even faster pulsars have been found since). Its location, however, was even more significant than its speed.

The Crab Nebula is actually the debris from a supernova explosion, one which was observed from Earth by Chinese astronomers in AD 1054. Walter Baade had pointed out, years before, that if supernova explosions left neutron stars behind, the best place to look for a neutron star would be in the middle of the Crab Nebula. He had even identified a particular star in the Crab Nebula that he said might be the neutron star left behind by the explosion. Until 1968, almost everybody else thought he was wrong—although, as the fact that neutron stars were even mentioned by Hewish's team in the pulsar discovery paper shows, by the middle of the 1960s a few theorists were dabbling with calculations of the structure and behaviour of such objects. But the radio observations showed that the Crab Pulsar seemed to be in the same place as the star that so interested Baade.

Further studies showed that this star was actually flicking on and off, in *visible* light, 30 times a second—something that nobody could have conceived as being possible just a few months before. A star that flickered so rapidly was beyond the wildest imaginings of the most daring theorist. Yet it did. It was indeed the pulsar, energetic enough to be detected in visible light, not just with lower energy radio waves.

By the time those observations were made, at the Steward Observatory on Kitt Peak, Arizona, in January 1969, everyone was convinced that pulsars are indeed neutron stars. And it had also become clear that, in spite of their name, they are not pulsating, but rotating, and beaming radio waves (and in some cases light) out through space from an active site on their surface. The pulses produced by a pulsar are the equivalent of a celestial lighthouse (but natural, not the product of an alien civilization), flicking its beam past the Earth repeatedly as the underlying star rotates. There is now an overwhelming weight of evidence that this is indeed the case, and that pulsars are neutron stars spinning so fast that with many of them a spot on the equator is being whirled around at a sizable fraction of the speed of light.

The person who put the idea down on paper, and published it in *Nature* in the early summer of that year, was Thomas Gold, who thereby gained fame as the man who worked out the true nature of pulsars. In fact, not long before the announcement of the discovery of pulsars (and after Jocelyn Bell had first noticed the bit of scruff on her charts), Franco Pacini had published a paper in *Nature* late in 1967 in which he pointed out that if an ordinary star did collapse to form a neutron star, the collapse would make it spin faster (like a spinning ice skater drawing in her arms) and strengthen the star's magnetic field, as it was squeezed, along with the matter, into a smaller volume. Such a rotating magnetic dipole, said Pacini, would pour out electromagnetic radiation, and this could explain details of the way the central part of the Crab Nebula still seems to be being pushed outwards, nearly a thousand years after those Chinese astronomers saw the supernova explode. It may seem a little unfair that Gold to some extent stole Pacini's thunder by linking the rotating neutron star idea with pulsars. It is worth mentioning, however, that similar ideas about the source of energy in the Crab Nebula had been aired by Soviet researchers a couple of years previously, and that as far back as 1951 Gold had speculated, at a conference held at University College, London, that intense radio noise might be generated in the neighbourhood of collapsed, dense stars.

One key feature of the application of these ideas to pulsars was predicted by Gold in his 1968 paper. Rotating neutron stars ought to slow down slightly, spinning less quickly as time passes. When measurements were carried out by a team using the 300-metre dish antenna built into a natural valley in Arecibo, Puerto Rico, the Crab Pulsar's pulse rate was indeed found to be slowing down, by about a millionth of a second per month. 'Gold's model' could no longer be doubted.

Jocelyn Bell got her Ph.D. (and Hewish later received a Nobel prize) chiefly for discovering pulsars. Since 1968, dozens of astronomers have built entire careers upon her discovery of that little piece of scruff.

ten

How Galaxies Form

▼

PROBABLY THE SINGLE MOST profound observation we have of the Universe in which we live is that it is a cold, dark place with a scattering of hot, bright stars. On the scale of the Universe itself—a cosmological scale—it makes more sense to regard the galaxies, rather than individual stars, as the fundamental 'units' of matter. But the dichotomy remains to be explained: how did the galaxies, containing thousands of millions of hot, bright stars, come to be scattered across the cold, dark backdrop of space? By no means is this a trivial problem. We are so used to the appearance of the night sky that our natural, human reaction is simply to accept that 'of course' the sky is dark and the stars are bright—just as 'of course' an apple falls down from a tree, not up. It took Isaac Newton to appreciate the significance of the apple's fall. Many modern cosmologists have puzzled over the significance of the dark night sky, and Professor Sir Herman Bondi has commented that 'the fact that our night sky is very black, with very bright points, the stars, in it, may be the profoundest piece of knowledge of the Universe that we have'.

The reason for this dramatic statement is that all known physical systems tend to reach a state of thermodynamic equilibrium. This means, in simple terms, that heat differences tend to average out. Place a lump of ice in a dish of hot water, and the ice melts while the water cools. Never do we observe a pan of warm water suddenly dividing up into a lump of ice and a boiling puddle. In the Universe at large, similar processes are at work, with stars pouring out energy into the cold darkness between them. So far, so good—the universe *is* 'tending towards thermodynamic equilibrium', as it should, given the state it is in today. But how did it ever get to such a state? The temperature difference is impressive; the only meaningful guide to the 'temperature' of the whole Universe is provided by the cosmic microwave background radiation, a faint hiss of radio noise detected from all directions in space. It is equivalent to the radiation from a 'black body' (the physicist's idealized example of an energy radiator) at a temperature close to 3 K, just about –270° C. But the temperature at the surface of a star like the Sun is several thousand K, and stellar interiors may be several million K. If the Universe started its life as a uniform ball of hot radiation, blasting outwards from a Big Bang, something must have happened to disturb the equilibrium long ago, and allow hot stars—or, rather, galaxies full of hot stars—to form in the expanding Universe. The hot Big Bang model is very firmly established, in the eyes of most cosmologists, as a good description of the real Universe, which seems to have begun its expansion from a hot fireball around 15,000 million years ago (various estimates range from 10,000 to 20,000 million years, which is a very good agreement by astronomical standards). The presence of the background radiation itself is today seen as clinching evidence for the reality of the Big Bang, the last traces of a fireball, once so hot that matter could not exist. All the energy of the Universe was in the form of radiation.

Galaxies themselves behave as they should in an expanding Big Bang universe, flying apart from one another like fragments of an exploding shell. Indeed, it was measurements of the velocities of galaxies, made by Edwin Hubble in the 1920s, which first suggested to cosmologists that the Universe is expanding. The measurements made by Hubble, and since refined by many others, make use of the way spectral lines in the light from distant galaxies are shifted towards the red end of the spectrum. This redshift is interpreted as a Doppler effect of motion (in this case, the outward motion of an expanding Universe), rather similar to the change in pitch of the siren on a fast-moving police car or ambulance as it passes you. It was only the discovery of the background radiation in the 1960s, however, that convinced cosmologists that the Universe did originate in a hot Big Bang. In 1978, Arno Penzias and Robert Wilson received a Nobel prize for the discovery of the 3 K background, which indicates the importance of the discovery as a turning point in the development of our ideas about the Universe. In the decade and a half since their discovery, with the aid of measurements of the background, other cosmologists have been able to build up a picture of events covering the very birth pangs of the universe as we know it, a picture graphically described by another Nobel prizewinner, Steven Weinberg, in his book *The First Three Minutes*.

By the mid-1970s, astrophysicists and cosmologists were in a curious position. They could explain the behaviour of the fireball stage of the Universe, its first few minutes of life, in great detail. They could also explain the lifecycle of a star like the Sun, in an established galaxy of stars (our Milky Way) in a Universe thousands of millions of years old. But they still could not explain in satisfactory detail how galaxies of stars formed in the expanding Universe as the fireball cooled. How did hot matter get segregated out from the uniformity of the fireball, while the radiation cooled right down to 3 K?

This had always been a problem, even in the decades before the discovery of the background radiation confirmed the validity of the Big Bang description of the Universe. In a perfectly smooth expanding Universe, filled with the mixture of 75 per cent hydrogen and 25 per cent helium gas that was the material relic of the Big Bang itself, it is very difficult to make galaxies and stars. The expanding Universe spreads the gas thinner, yet must pull local concentrations of gas into collapsing, dense clouds from which stars and galaxies could be born. Since Hubble's time, it had been clear that even if irregularities did develop in the gas filling an expanding Universe—whirlpools and eddies on a cosmic scale—it would take a very long time for the inward pull of gravity to overcome the thinning effect of expansion and form irregularities on the scale of galaxies. Indeed, right up until the 1970s the simplest calculations of the time it would take for a cloud big enough to form a galaxy to collapse in the expanding Universe indicated an interval longer than the age of the Universe itself! This uncomfortable state of affairs was one of the reasons for the rise of the steady state theory as a challenger to the Big Bang in the 1950s. Some cosmologists argued then that there may never have been a 'beginning', even though the Universe had always been expanding. They suggested that new matter was always being created to fill the gaps between galaxies in the expanding Universe and that the Universe was infinitely old. This allowed as much time as you would like for the new gas between galaxies to condense out into new galaxies!

The discovery of the background radiation killed off the steady state theory by offering compelling evidence of the reality of the Big Bang. But this also left galaxy formation as a dilemma, although cosmologists were too excited by the physical reality of the background radiation and the new tool for improving their models of the Big Bang itself to worry overmuch about that at the time. It took, in fact, about ten years for the implications of the 3 K radiation to be fully assimilated into cosmological thinking, and then a new generation of astrophysicists began to turn their attention to the puzzle of galaxy formation—but armed now with a detailed knowledge of the fireball in which the Universe was born and out of which galaxies must have been created. At the same time, ever-improving observational techniques with ever better telescopes and other instruments were able at last to provide evidence for or against the developing new theories on a scale covering a sizeable fraction of the universe. The result of this two-pronged attack, from theorists and observers, was that by the end of the 1970s a new theory of galaxy formation emerged. This was a landmark in understanding the relationship between galaxies and the Universe. First, the entire process of galaxy formation can now be explained within the timescale of the age of the Universe, from the improved understanding of the way fluctuations could have grown in the last stages of the fireball. Secondly, the new theory suggests that every bright galaxy we see may be surrounded by a 'superhalo' of dead stars containing perhaps as much as ten times more matter than in the bright stars. The visible part of a galaxy may be like the tip of an iceberg, with the bulk of its matter hidden. That implies that 90 per cent or more of the matter in the Universe really is cold and dark, in thermodynamic equilibrium with the cold, dark Universe itself, just as thermodynamics suggests things 'ought' to be.

There is direct evidence that the bright matter in the Universe makes up only 10 per cent of the mass. Studies of how galaxies rotate—which depend once again, on the ubiquitous Doppler shift—give us this information. While the redshifts in the spectra of light from distant galaxies tell us the recession velocities of those galaxies, galaxies that are closer to us can be studied in more detail, and the variations in the spectra of light from different parts of one galaxy reveal how the stars in those different parts are moving relative to one another. In particular, this analysis of the velocity Doppler shift in different parts of the galaxies shows how the galaxies rotate. The observations are by no means easy, and they represent a triumph of modern astronomy in their own right. What they reveal is certainly worth the effort, since the 'rotation curves' derived from these studies clearly show the dragging effect of much greater masses surrounding the visible bright galaxies. A galaxy rotating on its own would spin differently because it would be free from the gravitational field of the dark matter. When the optical observations are extended by radio observations at 21 centimetres (the wavelength at which hydrogen produces a strong 'line'), the same results are found. These superbly accurate new measurements were coming in, in the mid-1970s, and as they did they provided the direct stimulus for the theorists, armed with their new understanding of the fireball, to come up with a satisfactory model of galaxy formation.

The story theorists now have to tell begins with fluctuations in the last period of the fireball era, when the Universe was filled with hot matter and hot radiation. The matter was what we would call an ionized gas, with negatively charged electrons and positively charged nuclei (chiefly single protons), still able to interact with radiation and not bound into electrically neutral atoms. The constant interactions between the charged particles and the radiation kept the two forms of energy distributed smoothly, and the smoothness of the cosmic background radiation today, which has scarcely interacted with matter since that time, is a sign of the smoothness of the Universe then. But there must have been some fluctuations from perfect uniformity among the particles of the ionized gas. As they moved about at random, it must have happened that from time to time one spot would get more than its share, temporarily increasing in density, while somewhere else there would be a temporary deficit of particles, thinning out the cosmic soup. And radiation too, under these conditions, is subject to the same kind of fluctuation in energy density. Uniform the Universe certainly was, over large spans of space and reasonable stretches of time. But it must have been bubbling with activity as, first in one place and then in another, denser or thinner pockets of mass or energy formed, only to be dissipated by matter-radiation interactions and reformed into new patterns.

The pattern of bright galaxies we see today is left over from the very last fluctuations of the fireball, the pattern that became frozen in as the Universe cooled to the point where nuclei claimed their quota of electrons to form electrically neutral atoms, and radiation and matter decoupled, leaving gravity as the dominant universal force.

In a mixture of ionized gas and radiation, there are two kinds of density change which can occur. If it is just a question of particles getting together briefly in a patch of increased density, then it is called an *isothermal* fluctuation (because there is no change in energy density, which corresponds to no change in temperature). If, however, the fluctuation increases the local density of both matter and radiation, it is called *adiabatic*. Both kinds of fluctuation must have occurred in the fireball. But the pattern of galaxies left over today very clearly shows that isothermal fluctuations dominated at the end.

Rather than trying to calculate the later evolution of a Universe filled with both kinds of fluctuation, astronomers have looked at each kind, and its consequences, separately. Purely for historical reasons—because some very significant pioneering work on the subject was carried out by Academician Yakov Zel'dovich—Russian theorists have tended to concentrate on the growth of adiabatic changes in an expanding Universe, while Western theorists have looked in more detail at isothermal changes.

There is a crucial difference between the two kinds of change. Isothermal fluctuations (matter only) can be of any size, but adiabatic fluctuations (matter plus radiative energy) can survive and grow only if they start bigger than a critical size. Small fluctuations are quickly smoothed away and damped out by the changes in the energy density of the radiation. If isothermal changes dominated in the era of recombination, the universe should be filled with many relatively small galaxies, clustered together into larger groups, which themselves cluster into superclusters, and so on in a continuing hierarchy. But if adiabatic changes dominated in the real Universe, then the first aggregations of matter must have been many times more massive than even a cluster of galaxies as we see it today. The original aggregations of matter would have been huge gas clouds which then collapsed under gravity and broke up into smaller clouds, which in turn broke up into individual galaxies and stars.

So the adiabatic fluctuations produce not a hierarchy of clusters (like a set of nested Russian dolls), but a uniform pattern of very similar superclusters side by side (like a box of toy soldiers).

To analyse fully the clustering of galaxies in the real universe requires a lot of patience and a good, fast computer to do the calculations. More than a million galaxies have now been located as members of one cluster or another, and the statistical calculations show very strong support for the isothermal (hierarchical) model. At the same time that these calculations were being made in the late 1970s, improved calculations of the physical conditions at the time when nuclei and electrons recombined were also being made, drawing on the improved cosmology of the 1970s which had been built on the discovery of the cosmic background radiation. These estimates suggested that, in fact, any adiabatic fluctuations forming in the era of recombination of atoms would be vastly bigger than even a supercluster of galaxies, and so stable that they would remain forever as huge gas clouds, never fragmenting at all into galaxies as we know them! The distribution of matter in the Universe today results from isothermal changes in the distribution of matter (local density fluctuations in the last phase of the cosmic fireball, during recombination).

So far, so good. But how did galaxies like our own actually form from the primordial fluctuations? The complete story has still to be unravelled. But the outlines at least are now becoming clear. And the development of this new picture involves a major upheaval in astronomical thought. A wealth of new evidence is telling us that many of our old ideas about galaxies (and by 'old' I mean pre-1975) are quite simply wrong. One of the teams at the cutting edge of this new attack on the problem of galaxy formation is at the Institute of Astronomy in Cambridge, England, under the leadership of Professor Martin Rees.

The first coherent attempt to produce an overall picture of the new understanding of galaxies is based on the reasonable argument that just after recombination, when the overall density of the Universe was much higher than it is today, conditions were ideal for star formation, and the first stars formed then in great numbers. Quite simply, when the Universe as a whole was denser, it was easier for local condensations to become dominated by their own gravitational attraction and collapse further into stars. These stars may have been very small, something intermediate between our Sun and the giant planet Jupiter. If so, they never released much in the way of nuclear energy and never burned very bright. They may have been very big, however—superstars perhaps containing the matter of a million Suns—and they burnt their nuclear fuel quickly. As they died they scattered the 'ashes' in the form of heavier elements built up from hydrogen and helium by nuclear fusion and left behind burn-out cinders, neutron stars or black holes. Either way, the dark stars left behind would have been distributed through the hierarchical clustering that was the heritage of the original isothermal fluctuations. And only 10 per cent of the original hydrogen and helium gas was left to cool in the space between these early stars.

As the Universe continued to expand, moving the clusters of now dead stars apart from one another, the tiny remnant of old, cold gas must have sunk into the middle of each supergalaxy, sliding down the gravitational potential well and collapsing to form the bright galaxies that we see today, embedded deep within the real galaxies of dark, cold stars.

Apart from the fact of their existence, the most striking feature of galaxies is the way they divide into two classes, the flattened, disk-like spirals and the fatter, rounder ellipticals. This, too, has been a puzzle for astronomers since Hubble's day, but it can be explained within the framework of our developing understanding of galaxy formation. The slowy rotating spirals, like our Milky Way, can be explained very simply as the natural products of a gas cloud collapsing at the heart of a super-galaxy of dark stars. And it now looks very much as if not only ellipticals but the variety of peculiar galaxies seen in the Universe can be explained by interactions between two or more spirals once they have formed—interactions which may strip away spiral arms by tidal effects, merge two galaxies into one as one 'overtakes' another, or blast a whole galaxy to bits in a head-on collision. This suggestion was dramatically born out in the 1980s and 1990s, when improved telescopic observations of galaxies deep in the Universe (including pictures from the Hubble Space Telescope) showed exactly the kind of mergers and interactions between spiral galaxies going on that are required to turn many of them into ellipticals.

It is a notable achievement to have, at long last, a reliable framework, in the form of an understanding of the basics of galaxy formation in the expanding Universe, on which to build. So far, this has been a quiet revolution in astronomy, gaining none of the popular attention of stories such as the discovery of the background radiation, or pulsars, or of new theories such as the latest ideas about black holes. In its quiet way, however, the modern view of galaxy formation is as profound a change in our thinking as any of these. It explains old problems rather than creating new ones. This may be why it seems unspectacular in some ways. But by bringing our view of the Universe more closely in step with the requirements of thermodynamic equilibrium, it has relegated us to a place not just on a small planet circling an insignificant star in an unspectacular galaxy, but with that whole galaxy, like all the other bright matter in the dark Universe, merely the leftover embers of the glory of the hot fireball, flickering amongst the ash of dead cold matter ten times as extensive as the bright stars. Along the way, it has required the development of new ideas hand in hand with new observations, both equally profound and neither making much sense without the other, a classic example of the indivisibility of modern astronomy.

eleven

The Man Who Proved Einstein Was Right

▼

ALBERT EINSTEIN HAS BECOME an almost mythical figure, part of the folklore of our time. He is the archetypal genius, the white-haired, slightly eccentric but amiable old man who pierced to the heart of complex problems by applying an almost childlike naivety, and by asking questions so obvious that no one else had thought to ask them. Much of this is true, just as it is true that he was not thought to have been particularly bright in school, made no pronounced impact on the academic world as a student, and had to work as a technical expert in a patent office in the early years of this century while he developed three major new ideas in physics in his spare time. In one respect, however, the stereotypical image doesn't tell us the truth about the man who made these revolutionary contributions to science.

In the early 1900s, Einstein was not a white-haired, genial patriarch who dressed for comfort rather than elegance and sometimes didn't bother to wear socks. As pictures from this period show, he was a dark-haired, handsome young man who dressed with conventional smartness. This is important, for Einstein's greatest ideas were *youthful* ideas. They provided new insights, overturned established wisdoms and were truly revolutionary. The burst of activity that brought Einstein to the wide attention of the scientific community was completed in 1905, when he was twenty-six. His greatest achievement, the General Theory of Relativity, was published just ten years later. And although he lived until 1955, becoming the genial old professor of folklore, his greatest works were all behind him before the end of the First World War. Science, especially its mathematical side, is like that: only young minds can stretch to discover and embrace new concepts, and if the new concepts are as dramatically different from old concepts as those Einstein developed, it can take the rest of your life, or several lifetimes, to work out the implications.

Einstein, however, had one great stroke of good fortune. His exotic new mathematical theory was understood, shortly after publication, by another great scientist of the twentieth century, and this astronomer realized that an almost unique opportunity to test the theory was about to occur. It was that successful test of General Relativity that thrust Einstein into the public eye and gave us, as a direct result, the public persona, the archetypal image, of the grand old man of science that is remembered with affection to this day. The man who proved Einstein was right also gave us, in a very real sense, the Albert Einstein we all know and love.

General Relativity is, above all else, a theory of gravity. It predicted, among other things, that light can be deflected by gravity. The best way to understand how this bending of light occurs is to cast aside our preconceived ideas about force and space, and to take up the ideas presented by Einstein, initially in 1915 and in a complete form in 1916. These ideas envisage what we think of in everyday terms as empty space, as something almost tangible, a 'spacetime' continuum in four dimensions (three of space and one of time) which can be bent and distorted by the presence of material objects. It is those bends and distortions which provide the 'force' of gravity.

Forget about the four dimensions of spacetime for a moment, and think of a two-dimensional elastic surface. Imagine a rubber sheet stretched tightly across a frame and making a flat surface. This is a 'model' of Einstein's version of empty space. Now imagine dumping a heavy bowling ball in the middle of the sheet. It bends. This is Einstein's 'model' of the way space is distorted near a large lump of matter. When you roll marbles across the flat rubber sheet, they travel in straight lines. But when the sheet is distorted by the bowling ball, any marble you roll near the ball follows a curved trajectory around the depression in the rubber sheet. This, said Einstein, in effect, is where the 'force' of gravity comes from. There isn't really any force; objects are simply following a path of least resistance, the equivalent of a straight line, through a curved portion of space, or spacetime. The object can be a marble, a planet or a beam of light; the effect is the same. When it moves near a large mass—through a gravitational field—its path gets bent.

General Relativity predicted exactly how much a beam of light should bend when it passes near the Sun. The mathematics may be esoteric and the concepts, such as bent space, bizarre, but Einstein's General Theory made a clearcut and testable prediction. It appeared in 1916, in the middle of the First World War, when Einstein was working in Germany. The British astronomer Arthur Eddington, in a nation at war with Einstein's homeland, learned of the new theory and its prediction from a colleague in neutral Holland. This German prediction was confirmed by a British observation made in 1919, while the two countries were still technically at war, having signed an armistice but not yet a peace treaty. Partly for these reasons, it caught the popular imagination like no other discovery in the physical sciences, causing a stir comparable only to the response to the publication of Darwin's ideas on evolution in the previous century.

A scientific theory can never, strictly speaking, be *proved* correct. The best any theorist can hope for is that his or her theory will make a prediction which can later be tested and found to be accurate, to within the limits of observational or experimental error. In this sense, Einstein's theory has proved to be a more complete theory than Newton's theory of gravity, producing predictions which are more closely in agreement with observations. This is the special, restricted sense in which Einstein's theory was 'proved' right in 1919. And the man chiefly responsible for obtaining the proof was Arthur Eddington.

Eddington was three years younger than Einstein, having been born in 1882 in Kendal, Cumbria—the home of the famous 'mint cake' carried by mountaineers up Everest as part of their iron rations. His father died in 1884, and the young Eddington moved with his mother and sister to Weston-super-Mare, a seaside town in Somerset, where he was brought up and attended the local school. He was a Quaker throughout his life—something that was to be important to the confirmation of the Einsteinian prediction of light bending, in a roundabout sort of way—and an outstanding scholar who went first to Owens College in Manchester (the college which became the University of Manchester) and then, after graduating in 1902, to Cambridge. Three years later he graduated from the University of Cambridge, and after a short spell of teaching in 1907 he became a Fellow of Trinity College. He also took up a post at the Royal Greenwich Observatory, as Chief Assistant. In 1912, at the age of twenty-nine, he became Plumian Professor of Astronomy and Experimental Philosophy at the University of Cambridge, and in 1914 the Director of the Cambridge Observatories.

If all that makes him sound a formidable man, the impression would be only half-correct. Eddington was also a brilliant communicator, who became one of the leading popularizers of science in the 1920s and 1930s, and he had a well-developed sense of humour and of the bizarre. In later life, he told how one of his schoolboy games was to make up sentences which obeyed all the rules of English grammar, but made no sense—one example was 'To stand by the hedge and sound like a turnip.' In his writings on theories such as quantum physics and relativity, he was prone to slip in a bit of Lewis Carroll to help get a point across. There certainly has to be something out of the ordinary about a man who can begin a chapter of a book entitled *Philosophy of Physical Science* with the sentence:

> I believe there are 15, 747, 724, 136, 275, 002, 577, 605, 653, 961, 181, 555, 468, 044, 717, 914, 527, 116, 709, 366, 231, 425, 076, 185, 631, 031, 296 protons in the Universe, and the same number of electrons.

Perhaps even more remarkably, the reasons why Eddington came up with this large number are still of interest to cosmologists (though that is another story).

Eddington will be remembered for two great achievements. As much as anyone else, he invented the subject of astrophysics—the study of how physical laws deduced here on Earth, together with observations of the light from stars, can explain how the processes going on inside stars keep them hot, and how the stars must change as they age. He was also the definitive popularizer of Einstein's Theories of Relativity in the English language, not just in the sense of communicating these ideas to lay persons, but also as the scientific interpreter who made them clear to his colleagues, and wrote textbooks on the subject which helped spread its message. Fascinating though all of Eddington's life and work was, the one thing that stands above all the rest is his response to the prediction that light must bend when it passes near the Sun.

Einstein's first announcements of the General Theory were communicated to the Berlin Academy of Sciences during the latter half of 1915, and published in more detailed form the following year in the journal *Annalen der Physik*. This paper is called 'The foundation of the General Theory of Relativity', and is widely accepted as the greatest Einstein ever wrote. Copies of Einstein's papers went, naturally enough, to his friends in the neutral Netherlands. One of those friends, Willem de Sitter, sent copies of Einstein's papers to Eddington. In 1916 and 1917, de Sitter also sent three of his own papers to the Royal Astronomical Society for publication. These were partly reviews of Einstein's work, explaining its significance, but in the last of the three de Sitter also presented for the first time a description, based on General Relativity, which required the Universe to be expanding. Eddington was Secretary of the Royal Astronomical Society at the time, and we know that he read the papers carefully and reported on them to the Society's meetings, prior to their publication. The one person who had the intellectual ability and background to appreciate fully the significance of Einstein's new work was in exactly the right place, at the right time, to get the news. Fate had several more twists to add to the story before Einstein's new theory was proved correct.

The way to test for light bending, as Einstein pointed out, was to look at stars seen near the Sun during an eclipse. Normally, of course, the bright light of the Sun makes it impossible to see stars in that part of the sky, but with the

Sun's light temporarily blotted out by the Moon it would be possible to photograph the positions of stars which lie far beyond the Sun but in the same direction in the sky. By comparing such photographs with photographs of the same part of the sky made six months earlier or later, when the Sun was on the other side of the Earth, it would be possible to see any shift in the apparent positions of the stars produced by the light bending effect. What the astronomers needed was an eclipse of the Sun. Ideally, if they could have chosen the eclipse they wanted, they would have asked for one on 29 May, in any year, because just then the Sun is seen passing in front of an exceptionally rich field of bright stars, in the direction of the Hyades. Eclipses are quite frequently visible from some part of the earth, but an eclipse on 29 May (or any other particular day of the year) is something that happens only very rarely. As Eddington himself commented, 'it might have been necessary to wait some thousands of years for a total eclipse of the Sun to happen on the lucky date'. But by a remarkable stroke of good fortune, there was an eclipse due in 1919—on 29 May. It was too good an opportunity to miss, provided the war was over in time to organize an expedition to observe the eclipse, which would be visible from Brazil and from the island of Principe off the west coast of Africa.

In 1917, the plot began to thicken. The Astronomer Royal, Sir Frank Dyson, was enthusiastically in favour of organizing two expeditions to observe the 1919 eclipse, and contingency plans began to be made. Meanwhile, conscription was introduced in Britain, with all able-bodied men eligible for the draft. Eddington was thirty-four and able-bodied; he was also a devout Quaker and a conscientious objector. This was a difficult thing to be in 1917, and the position was further complicated by the recognition of the scientific community that Eddington was a scientist of the first rank. The physics community still felt deeply the loss of

Henry Moseley, a pioneering X-ray crystallographer killed in action at Gallipoli in 1915, and questioned the wisdom of the government in sending the best scientists of the day to die, perhaps, in the trenches. A group of eminent scientists pressed the Home Office to give Eddington an exemption, on the grounds that Britain's long-term interests would be best served by keeping him at his proper work. The Home Office eventually agreed and wrote to Eddington, sending a letter for him to sign and return. Eddington added, however, a footnote to the letter to the effect that if he were not deferred on the stated grounds he would claim deferment on the grounds of conscience anyway.

It was an honest and principled stand, which left the Home Office with a problem, and the scientists who had pleaded on Eddington's behalf more than a little upset. The law of the time stated that a conscientious objector must perform uncongenial work in agriculture or industry. Eddington was quite prepared to go and join his Quaker friends. The upshot of a further round of debate, involving Dyson, as the Astronomer Royal, was that Eddington's draft was deferred, but with the 'condition' that if the war ended by May of 1919 he *must* lead an expedition to test the light bending prediction of Einstein's theory! (There are several versions of this story which give a slightly different emphasis but report the same series of events. I have followed the account given by Subrahmanyan Chandrasekhar in his book *Eddington*, which is also the source of the direct quotations from and about Eddington given here.)

Eddington had led an expedition to Brazil to study the 1912 eclipse of the Sun. He needed all of his experience to ensure the success of his part of the twin 1919 expeditions, which were planned throughout 1918. The plan was that Eddington and a Cambridge team would go to Principe, while Dyson would organize a team from the Royal Observatory, Greenwich, to observe the eclipse from Brazil. Unfortunately, no work could be done by the instrument-makers until the Armistice was signed; they were too busy building weapons of war. But the expeditions had to sail in February 1919. The Armistice was signed, of course, on 11 November 1918, and the anniversary of that date is still marked by memorial services. In a few hectic weeks everything was made ready, and the expeditions set off.

The Brazil expedition had perfect weather for the occasion and obtained a series of excellent photograph plates of the star-field around the Sun at the time of the eclipse. But for logistic reasons these plates were not processed and studied immediately. On Principe, Eddington waited anxiously as the appointed day dawned rainy with a cloud-covered sky. More in hope than expectation, all the arrangements to photograph the eclipse were made, and just near the time of totality the Sun showed dimly and the plates were exposed. The result was just two plates showing the stars needed for the test. Eddington had arranged for these plates to be examined on the spot, 'not entirely from impatience,' as he put it, 'but as a precaution against mishap on the way home'. One of the successful plates was duly developed and analysed on Principe, and Eddington compared it with another plate of the same part of the sky that he had brought with him. The measurements required were simple. Three days after the eclipse, Eddington knew that he held in his hands the proof that Einstein's General Theory of Relativity was correct.

The full analysis of the eclipse observations took several months, so definite news that his prediction had been confirmed did not reach Einstein until September of 1919. The full results of the expeditions were announced in London to a packed joint meeting of the Royal Society and the Royal Astronomical Society on 6 November 1919, and they produced a wave of publicity in a world eager for news of anything except war. The headlines read 'light does not go straight', 'Revolution in science', 'Newtonian ideas overthrown' and 'Space "Warped'". Einstein was established in the public eye as the great scientist of the twentieth century, perhaps of all time. The General Theory of Relativity was accepted as the greatest scientific theory of all time.

There were other tests of Einstein's theory. It had already explained a tiny variation in the orbit of Mercury around the Sun that is not predicted by Newton's theory of gravity, and this convinced astronomers that Einstein's theory worked better than Newton's. Other eclipse expeditions followed, and these tests have since been repeated many times, often far more accurately than Eddington's first analysis of his plates on Principe. Totally different tests of General Relativity, involving redshifts caused by gravity in the light from stars and subtle changes in the radiation from pulsars (undreamed of in 1919) all point to the same conclusion. Yet, whatever tests have been carried out since, 29 May 1919 stands as the day when science made the observations that proved Einstein correct, and 6 November 1919 stands as the day the public were made aware of the fact.

Eddington, however, would not have liked us to leave the story on quite this note. Following the two expeditions of 1919, the next eclipse study was in 1922, when a Lick Observatory expedition went to Wallal in Western Australia and provided the third measurement of the Einsteinian light bending effect, and again showed the results predicted by General Relativity. The news was announced to the April 1923 meeting of the Royal Astronomical Society where Eddington, characteristically, quoted Lewis Carroll: I think that it was Bellman in *The Hunting of the Snark* who laid down the rule 'When I say it three times, it is right.' The stars have now said it three times to three separate expeditions; and I am convinced their answer is right.

And that is *exactly* the appropriate note on which to end my tale of the man who proved Einstein was right!

twelve

The Man Who Invented Black Holes

▼

BLACK HOLES ARE PRODUCTS of gravity. Modern science began with Isaac Newton, who, among other things, developed the first scientific theory of gravity, a little over three hundred years ago. For the first time, scientists were able, using Newton's laws, to explain the motion of heavenly bodies by the same principles that apply to the behaviour of objects on Earth. In the famous analogy, both the fall of an apple from a tree and the orbit of the Moon about the Earth could be explained by the same set of equations. Newton's description of gravity, of course, was later incorporated within Albert Einstein's General Theory of Relativity, and black holes are generally regarded, rightly, as essentially relativistic objects. But it is some indication of the power of Newton's own theory that less than a hundred years after the publication of his epic volume, the *Philosophiae Naturalis Principia Mathematica*, generally regarded as the most important single book ever published in physics, and equally generally referred to simply as the *Principia*, Newtonian gravitational theory had already been used to describe what we would now call black holes. Indeed, the surprise is that Newton, who investigated the nature of light as well as gravity, did not realize that his equations suggest the existence of dark stars in the Universe, objects from which light cannot escape because gravity overwhelms it. Newton's *Principia* contains the heart of what is known as classical mechanics—the three laws of motion, and a theory of gravity. The shoulders on which he could be said to have stood, metaphorically, in developing these ideas were those of Johannes Kepler, a German astronomer who published in 1609 the first two laws of planetary motion that now bear his name. Kepler developed those laws using tables of planetary positions painstakingly compiled by Tycho Brahe, a Dane who had settled in Prague, where Kepler became his assistant. Brahe died in 1601.

Kepler's first and second laws tell us that the orbits of the planets around the Sun are ellipses, not circles, and that a line joining a planet to the Sun traces out equal areas in equal times, wherever the planet is in its orbit. In other words, each planet moves faster when it is closest to the Sun, tracing out a short, fat triangle at one end of the ellipse, and slower when it is farthest from the Sun, tracing out a long, thin triangle at the other end of the ellipse. A third law, published several years later, relates the orbital period of each planet to the diameter of its orbit by a mathematical formula.

All this was intriguing and puzzling for seventeenth-century scientists, who searched unsuccessfully for an underlying explanation of Kepler's laws. We can only imagine the surprise Edmond Halley must have felt in August 1684, when he visited Newton in Cambridge. On mentioning that he was interested in the problem of orbital motion, Halley was told by Newton that he had solved that puzzle years ago. Despite his surprise, Halley kept his wits about him.

He persuaded Newton that this was such a significant discovery it simply had to be published, and just three months later Newton sent Halley a short paper on the subject. But this was not enough. Once Newton decided to publish his ideas, he began revising and rewriting this short paper until it grew into his great book, published (largely at Halley's expense) in Latin in 1687. It wasn't published in English until 1729, two years after Newton died.

Before Newton, scientists accepted the Aristotelian idea that the 'natural' state of an object is to be at rest and to move only when a force is applied to it. Newton realized that this only seems to be the case because we live on the surface of a planet, where things are held down by gravity. His first law says that every object (scientists usually use the term 'body') continues in a state of rest or of uniform motion in a straight line unless a force acts upon it. His second law states that the acceleration of a body (the rate at which its velocity changes, which means that either its speed or direction of motion changes) is proportional to the force acting upon it. And his third law says that whenever a force is applied to an object, there is an equal and opposite reaction force. If I push a pencil across my desk, for example, or press down on the desk top itself, I can feel the reaction to the force I am applying, a force pushing back on my fingertip. Although you might think, according to the second law, that the force of gravity should be making us accelerate towards the centre of the Earth, as long as we are standing on solid ground, the force of our weight pressing down is countered by an equal and opposite reaction force pushing up. The two forces cancel out, so there is no acceleration—unless you fall over or jump out of the window. If that happens, what hurts when you hit the ground is not the force of gravity but the reaction force of the ground, cancelling out the force of gravity and stopping your motion.

Using his own three laws and Kepler's laws, Newton explained the motion of the planets around the Sun, and of Jupiter's moons around Jupiter, as the result of a force of gravity which is proportional to one over the square of the distance between the Sun and a planet, or between Jupiter and a moon. This is the famous inverse square law. So, for example, when a planet is closest to the Sun, the force it feels is stronger, and it moves more quickly as a result. What's more, Newton said that this is not a special law that applies only to the orbits of planets around the Sun, but a universal law which describes the effect of gravity on everything in the Universe. The neatest example of this is one that Newton himself gave.

Newton knew that the acceleration caused by gravity near the surface of the Earth will make any body (an apple, for example) fall through a distance of 16 feet in the first second of its fall. (I'll use old-fashioned feet and inches in this example, because those are the units Newton used.) The Moon is 60 times farther from the centre of the Earth than the distance from the centre of the Earth to the Earth's surface, and according to Newton's first law, the Moon would 'like' to travel in a straight line at a constant speed—that is, at constant velocity. Even if the speed stays the same, any deviation from that straight line must be due to a force deflecting the Moon. According to the inverse square law, the force of the Earth's gravity acting at the distance of the Moon should be less than the force at the surface of the Earth by a factor of 60 squared, which is 3,600. So in one second the Earth's gravity should make the Moon shift sideways by a distance given by dividing 16 feet by 3,600. This works out to a little more than one-twentieth of an inch. For an object travelling at the speed of the Moon, at the distance of the Moon from the Earth, a sideways nudge of exactly this size, every second, is exactly enough to make it travel in a closed orbit around the Earth, completing one circuit every month.

The force of gravity acting on an object falling to the Earth acts as if all the mass of the Earth were concentrated in a point at the centre of the planet. The distance pertinent to the inverse square law is actually the distance between the centres of the two bodies involved: Sun and a planet, the Earth and a falling human being, or whatever. Newton proved this, but the first really accurate measurements that provide a measure of the gravitational constant were not made until the 1790s, more than a hundred years after the publication of the *Principia*. Henry Cavendish, a British physicist, made these measurements.

Although Newton lacked this laboratory proof of his law of gravity, he did, of course, believe that the law must apply everywhere and for all bodies. Since his other great achievement involved the study of light and an explanation of its behaviour in terms of tiny particles, 'corpuscles', streaming out from a light source and being reflected by mirrors or refracted by prisms and lenses, this makes it all the more remarkable that he seems never to have wondered how gravity would affect light. The first publication of any insight into that mystery had to wait almost a hundred years after the publication of the *Principia*.

The key to this idea, apart from Newton's law of gravity, was the measurement of the finite speed of light. It comes as a surprise to most people encountering these ideas for the first time that the speed of light was actually measured reasonably accurately *before* Newton published the *Principia*.

The calculation was made in the 1670s at the Paris Observatory by a Dane, Ole Römer, who was born in 1644. Among other things, Römer studied the behaviour of the moons of Jupiter, which were especially interesting to astronomers of the time because they represented a miniature version of the Solar System described by Copernicus and Kepler. The huge planet Jupiter is orbited by a set of moons in much the same way that the Sun is orbited by the planets.

Giovanni Cassini, one of Römer's more senior colleagues in Paris, had come to France in 1669, at the age of 44, to take charge of the new observatory. He became a French citizen in 1673 (changing his given name to 'Jean' at the same time). Cassini was a skilful observer, using the latest instruments in the new observatory. In 1675, he discovered the gap which divides the ring system of Saturn into two parts and is still known as Cassini's Division. His more important work, though, included studies of the behaviour of the satellites of Jupiter, and he carried out the first reasonably accurate measurement of the distance from the Earth to the Sun. It was these two sets of information that Römer put together to work out the speed of light.

One of the most obvious, and interesting, features of the behaviour of the moons of Jupiter is the way in which they are regularly eclipsed as they move into and out of the shadow of Jupiter itself. Even before he left Italy, Cassini had worked out a table of eclipses (rather like a bus timetable) for the four main satellites of Jupiter—Io, Europa, Ganymede and Callisto. Galileo had discovered them, using the first astronomical telescope, in January, 1610. With Kepler's laws, Cassini was able to forecast when the moons would be eclipsed. But Römer found that sometimes the eclipses were a little early and sometimes a little late, compared with the data in Cassini's tables. Concentrating on the behavior of Io, the innermost large moon of Jupiter, he found a regular pattern to this behaviour. The gap between eclipses was shorter than it ought to be when two successive eclipses were observed while the Earth was moving towards the position in its orbit closest to Jupiter (with both planets on the same side of the Sun) and longer when two eclipses were observed while the Earth was moving towards the position in its orbit farthest from Jupiter (on the opposite side of the Sun).

Even without knowing why this should be so, Römer

could make predictions based on the pattern he had discovered. In September 1679, he predicted that the eclipse of Io by Jupiter due on 9 November would be 10 minutes later than the standard orbital calculations suggested. The prediction was verified, and Römer stunned his colleagues by explaining the delay as being due to the finite time that it took light to cross space from Io to the Earth.

In the months leading up to that eclipse, the Earth had been moving in its orbit away from Jupiter. So when the previous eclipse had occurred, the light signalling that the eclipse had happened had not had quite so far to travel to reach the Earth. The November eclipse really had happened at the calculated time, said Römer, but by then the Earth was just farther enough away from Jupiter to require the light to take an extra 10 minutes to cross space to the telescopes of the Paris Observatory.

This was where Cassini's most important work, the investigation of the size of the Solar System, came in. In 1672, Cassini had carefully observed the position of Mars against the background stars from Paris, while his colleague Jean Richer made similar observations from Cayenne, on the north-east coast of South America. From these measurements, they were able to work out the geometry of an enormously tall, thin triangle, with a baseline stretching nearly 10,000 kilometres from Paris to Cayenne, and with Mars at its tip. This gave Cassini an estimate of the distance to Mars, from which he could work out the sizes of the orbits of other planets, including the Earth, using Kepler's laws and the time it takes each planet to travel around its orbit.

Cassini's estimate of the distance from the Earth to the Sun (now known as the astronomical unit, or AU) was 138 million km, by far the most accurate estimate made up to then. Tycho had come up with a figure of 8 million km, and Kepler himself made the distance about 24 million. Modern measurements indicate that the AU is actually 149,597,910 km. Using Cassini's estimate for the distance across the Earth's orbit to assess the extra distance light from the November, 1679 eclipse had to travel before reaching Römer's telescope, Römer calculated that the speed of light must be, in modern units, some 225,000 km per second. In fact using Römer's own calculation but with the modern estimate for the size of the Earth's orbit the figure is 298,000 km per second. The established value for the speed of light today is 299,792 km per second.

Whatever number actually comes out of the calculation, the real sensation surrounding Römer's work is the claim that the speed of light is finite. This meant that light signals do not travel instantaneously across the void of space, and that was such a dramatic claim that many scientists of the time refused to accept it. General acceptance that the speed of light is indeed finite came only after Römer's death, which occurred in 1710. Once you do accept that the speed of light is finite however, a new puzzle appears.

Anyone who has watched the launch of a Space Shuttle, even if only on television, is aware of the enormous effort that has to be expended to lift an object from the Earth's surface into a stable orbit around the Earth. Even more effort is needed to propel an object free of the Earth's gravity entirely and out through the Solar System, like the famous Voyager probes that sent back spectacular pictures from Jupiter and the other outer planets. The best way to measure the effort needed to break free from the Earth in this way is in terms of the speed of the escaping object. For any source of gravitation (which means any object in the Universe), there is a critical speed which has to be reached before an object leaving its surface vertically can escape. It is called the escape velocity. If you could magically make the Earth more dense, so that it contained more mass but stayed the same size, the escape velocity would increase. But although very large objects like the Sun and Jupiter contain much more mass than the Earth, this is spread out through a larger volume, so that the surface of the Sun or Jupiter is much farther from its centre than the surface of the Earth is from its own centre. Remember that gravity falls off as one over the square of distance from the centre of a body; this dilutes the strength of gravity and compensates for at least some of the extra mass. The escape velocity from the surface of a more massive (but larger) planet is not simply proportionately larger than the escape velocity from the surface of the Earth, but also depends on the density of the planet.

A rocket, like the Space Shuttle, builds up speed gradually as it uses fuel during takeoff. But we could achieve the same effect if we had a cannon powerful enough to fire cannonballs upwards at escape velocity. If we did this from the surface of the Earth, the cannonballs would have to leave the muzzle of the gun with a speed of 40,000 km per hour (or 11 km per second) in order to escape from the Earth's gravitational grip. Something moving with less initial speed will slow down, come to a halt and then fall back to Earth. Something moving faster than escape velocity will be slowed down, but never brought to a halt, and will continue moving out across space until it comes under the gravitational influence of another massive object. The escape velocity from the Moon is just 8,570 km per hour, and the escape velocity from Jupiter is nearly 220,000 km per hour, or just over 60 km per second.

In each case, the escape velocity is the speed with which cannonballs would have to be fired straight up to escape from the planet. What if we could mount our hypothetical cannon on the surface of the Sun? There the escape velocity would be more than 2 million km per hour—a speed which sounds truly impressive until you realize that it is only 624 km per second, which may be nearly fifty-seven times the escape velocity from the surface of the Earth but is still only

0.2 per cent of the speed of light. So light has no difficulty escaping from the surface of the Sun.

In the eighteenth century, scientists thought light consisted of corpuscles, in the way Newton had described them. They could be visualized as very much like tiny cannonballs fired from a glowing object. It was natural to guess that these corpuscles would be affected by gravity in just the same way as any other object. It was straightforward to work out the escape velocity from the Earth and to make a reasonable guess at the escape velocity from the Sun, assuming it had the same density as the Earth. But suppose there were objects in the Universe even bigger than the Sun. Suppose, indeed, that there were some stars so big that the escape velocity from their surface exceeded the speed of light. They would be invisible! This outrageous notion was suggested by John Michell in 1783, and caused a major stir in the sober company of the Fellows of the Royal Society.

Michell was born in 1724, and was seven years younger than his friend Henry Cavendish. At the height of his scientific career, he was regarded as second only to Cavendish in the ranks of English scientists, and today he is still known as the father of the science of seismology. He studied at the University of Cambridge, graduating in 1752, and his interest in earthquakes was stimulated by the disastrous seismic shock that struck Lisbon in 1755. Michell established that the damage had actually been caused by an earthquake centred underneath the Atlantic Ocean. He became Woodwardian Professor of Geology at Cambridge in 1762, a year after becoming a Bachelor of Divinity. In 1764, he became the rector of the parish of Thornhill, in Yorkshire, and some books give the impression that the Reverend John Michell was simply a country parson and some kind of dilettante and amateur scientist. In fact, his scientific reputation was well founded before he entered the Church, and he had already been elected a Fellow of the Royal Society (in 1760) before becoming a Bachelor of Divinity.

Michell made many contributions to astronomy, including the first realistic estimate of the distances to the stars and the suggestion that some pairs of stars seen in the night sky are not simply chance alignments of two objects at quite different distances along the line of sight, but really are 'binary stars' in orbit around each other. In spite of all this, Michell's name became almost forgotten in the nineteenth and twentieth centuries, to such an extent that in spite of a recent rehabilitation of his reputation, his brief entry in the *Encyclopaedia Britannica*, for example, does not even mention what now seems his most prescient and dramatic piece of work.

The first mention of 'dark stars' was made in a paper by Michell that was read to the Royal Society by Cavendish on 27 November 1783 and was published the following year in *Philosophical Transactions of the Royal Society*. This was an impressively detailed discussion of ways to work out the properties of stars, including their distances, sizes and masses, by measuring the gravitational effect of light emitted from their surfaces. Everything was based on the supposition that 'the particles of light' are 'attracted in the same manner as all other bodies with which we are acquainted,' because gravity is, said Michell, 'as far as we know, or have any reason to believe, an universal law of nature.' Among the many other detailed arguments in Michell's long-forgotten, but now famous, paper, he pointed out that

> If there should really exist in nature any bodies whose density is not less than that of the Sun, and whose diameters are more than 500 times the diameter of the Sun, since their light could not arrive at us . . . we could have no information from sight; yet, if any other

> luminiferous bodies should happen to revolve about them we might still perhaps from the motions of these revolving bodies infer the existence of the central ones with some degree of probability, as this might afford a clue to some of the apparent irregularities of the revolving bodies, which would not be easily explicable on any other hypothesis; but as the consequences of such a supposition are very obvious, and the consideration of them somewhat beside my present purpose, I shall not prosecute them any further.

What Michell had realized, in modern language is that a sphere 500 times bigger than the Sun (about as big across as the entire Solar System) and with the same density as the Sun would have a surface escape velocity greater than the speed of light. Although the idea stirred excited debate in London, as papers still in the files of the Royal Society show, it seems not to have spread outside England, for in 1796 Pierre Laplace, seemingly in complete ignorance of Michell's proposal, put forward essentially the same idea in his semi-popular book *Exposition du système du monde*.

Laplace's version of the dark star hypothesis—he called them 'corps obscurs', which translates as 'invisible bodies'—was essentially the same as Michell's. Unlike Michell, however, Laplace described his dark stars in terms of objects with the density of the Earth, which is greater than the density of the Sun, and he therefore calculated a diameter 250 times, rather than 500 times, that of the Sun. He suggested that there might exist

> in heavenly space invisible bodies as large, and perhaps in as great number, as the stars. A luminous star of the same density as the Earth, and whose diameter was 250 times greater than that of the Sun, would not, because of its attraction, allow any of its rays to arrive

at us. It is therefore possible that the largest luminous bodies of the Universe may, through this cause, be invisible.

Laplace's account of dark stars appeared in the first edition of the *Exposition*, published in 1796, and in the second edition, published in 1799. In 1801, the German astronomer Johann von Soldner calculated how a ray of light passing near a star would be bent by the influence of Newtonian gravity, and even speculated that the stars that make up the Milky Way might be orbiting a very massive, central 'corps obscur' of the kind proposed by Laplace (but he decided that they probably were not really doing so because he thought, incorrectly, that if they were, the sideways motion that would result ought to have been detected). In the edition of the *Exposition* published in 1808, however, and in all later editions, all reference to dark stars was deleted. Why Laplace abandoned the idea, nobody knows. But with Michell's work forgotten, the result was that the notion of black holes lay fallow for nearly 200 years, until revived, in the context of Einstein's theory of gravity, in the 1960s. The fact remains, though, that the groundwork had all been done by Newton, Michell and Laplace.

thirteen

Cosmic Gushers: White Holes and the Universe

▼

THE IDEA OF BLACK holes as ultimate sinks of matter and energy—perhaps even the ultimate 'death of the Universe'—has gripped the astronomical imagination in recent years, especially since X-ray astronomy has opened up a new window on the energetic processes going on in space. But it is less well known that for many years a small group of theorists has been investigating the reverse situation, in which matter and energy burst out into the Universe from a highly collapsed state, or singularity. The ultimate example of this kind of outburst is, of course, the explosive birth of the Universe itself from a singularity, in the initial 'hot Big Bang', and it is the possibility of the formation of the Universe in such an outburst that has encouraged the study of similar but smaller events—if you like, 'little bangs'. Among astrophysicists, these outbursts are generally called 'white holes', a name which conveys the idea of the opposite of a black hole. But since they are not holes at all, but fountains material gushing out into the Universe as we know it, I prefer to refer to them as cosmic gushers. And just as the death of the Universe, and of individual stars and galaxies, may lie in a black hole collapse, so the birth of the Universe, and of individual stars and galaxies, may be associated with cosmic gushers. In these days when tales of gloom and doom have become so fashionable, let us try to provide a cheering antidote by looking instead at how the cosmic gushers of white holes bring light and life to our Universe.

Some of the latest work on the theory of relativistic outburst has come from the fertile brain of Professor Jayant Narlikar, former colleague of Sir Fred Hoyle and one of the staunchest proponents of the steady state theory in the 1960s. Narlikar says of white hole theory today [1976] that

> it seems we are in a somewhat similar situation as that with regard to the existence of black holes about a decade ago. Then there were objections to supermassive objects forming on the grounds of angular momentum and fragmentation difficulties. In spite of these objections gravitational collapse of supermassive objects was studied . . . the evidence of exploding objects in the Universe indicates the necessity of studying white holes in spite of any possible stability problem seen at present.

So with the example of the rapid development of knowledge about black holes before us, and with the hope that the same ingenuity that overcame black hole problems will now be turned to those of cosmic gushers, this time we can get in on the ground floor of the newly developing field of study, with a likely grandstand view as it develops over the next few decades.

It is largely the work of Narlikar and his colleagues, published in 1975, that takes such studies firmly out of the realm of speculations and into the harsher world of astrophysical realities for the first time, and so those studies deserve pride of place in any current discussion. Cosmic gushers are useful in astrophysics because it is difficult to explain the observed, very energetic objects in the Universe in any other way. Within our own Galaxy, X-ray and gamma-ray sources are difficult enough to explain, and there is still no satisfactory alternative explanation for the 10^{55} joules of energy produced in violent outbursts from the nuclei of some galaxies, radio sources and quasars. All we can be sure of is that objects like Seyfert galaxies, radio galaxies and quasars are pouring matter outwards from very compact regions. That points the finger very provocatively at white holes.

Discussions about the exact nature and origin of a white hole read a bit like examples of the better kind of science fiction. But then, so do ideas about black holes in which time runs backwards, and even Einstein's Special Theory of Relativity has formed the basis of many a science fiction story, including *Planet of the Apes!* So the exotic image of a cosmic gusher is no reason alone to dismiss its existence as impossible.

Perhaps the simplest idea is that the Universe always was, in some sense, 'lumpy' right from the beginning, the Big Bang itself. The cosmic gushers we see today are therefore simply outbursts from regions that have been a bit late catching up with the Big Bang—retarded cores of expansion. This can be understood by imagining a Universe like ours, but running backwards and contracting towards a singular state. As the density increased, we would see regions of greater density—stars and galaxies—disappearing as they collapsed into black holes, with all the black holes eventually combining. Now run the 'film' the other way—the right way—to give the expanding Universe in which we live. As the expansion goes on, we will see regions of greater density popping out of singularities all around: white holes bursting into the visible Universe.

There are several other suggestions. The flow of our Universe through time could be like that of a river through space, with many branches joining it as it grows. Floating downstream, we would have no knowledge of these tributaries, with other galaxies and stars 'floating' down them, until they joined the main flow, bursting into our 'Universe'. Or perhaps, the most entertaining idea of all, a white hole could be literally the opposite of a black hole, at the other end of a 'tunnel' through spacetime. It pours back into the Universe all the material that has been swallowed up in a black hole somewhere else in time and space. It is easy to see why such speculations fascinate so many people, and there are plenty of questions still to be answered. So far, research such as that by Narlikar and his colleagues has concentrated one step along from the problem of the origin of these cosmic gushers. If we accept from the evidence of violent outburst in many astronomical objects that white holes may exist, even though we cannot yet put the finger on just where they come from, we can use relativity theory to predict what they should look like to observers like ourselves. The answer, according to the latest calculations, is that they should look very much like the energetic sources that have proved so difficult to explain in other ways—quasars, X-ray emitters and gamma-ray sources.

An exploding white hole will produce its own characteristic signature of electromagnetic radiation with a spectrum that can be predicted, but this basic spectrum will be modified by the blueshift induced by the violent movement of material towards an observer outside, just as our view of distant galaxies is affected by a redshift since they are moving away from us. Using the analogy of a 'little bang,' it is possible to apply the same Einstein—de Sitter equations which describe the overall expansion of the whole Universe to these outbursts, and in this way we can determine the general appearance of the spectrum to an observer—such as a terrestrial astronomer. Starting with the biggest events of this kind, it is possible to make out a reasonable case for white holes being responsible for the infrared and X-ray emission from Seyfert galaxy nuclei. That in itself is not too exciting; a 'reasonable case' seldom is. But no one is yet claiming that cosmic gushers can solve all the riddles of the Universe, and when we look closer to home, at events inside our own Milky Way Galaxy, things begin to look a whole lot more exciting. For example, the transient X-ray sources, discovered by X-ray satellites such as Uhuru, show a very sudden rise to peak energy, with a subsequent 'softening' of the spectrum, a power law with exponent in the range –1 to –3. Different detailed properties of these spectra can be explained in terms of the typical processes for the production of electromagnetic radiation: black-body, free—free or synchrotron radiation, as modified by the blueshift effect. And gamma-ray bursts look even more like classic cosmic gushers, with exponent –3 spectra softening with time exactly in line with the calculations. This development sets the study of white holes now firmly in the context of the present problems of high-energy astrophysics, and it opens the way for some intriguing ideas about one possible form of cosmic gusher to be taken seriously as a possible explanation for the origin of galaxies in our Universe in the form we see them.

These ideas are far from new. In the 1960s they provided the sort of exercise in juggling the equations of relativity theory that was thought to provide useful practice for a graduate student, but probably had no real relevance to the Universe in which we live. Such is the pace of modern astronomy that it is now possible to blow the dust off some of those calculations and present them as an up-to-the-minute contribution to the new debate about cosmic gushers! Apart from the fact that the Universe exists and expands, and leaving aside the observations of violent outbursts which hint that something dramatic goes on in the nuclei of many objects, there is evidence that our telescopes do not give us a complete picture of galaxies. Telescopic observations show that many galaxies must contain a great deal of invisible material in order to be gravitationally bound. This missing mass has lately been explained in terms of the possible presence of black holes, but of course that is merely the fashionable side of the coin. What is really required is the presence of matter in a collapsed form, and that could point to retarded cores as well as to black holes.

The average mass of a bright elliptical galaxy is some 800 billion solar masses, and the typical mass-to-light ratio is about 70. (If all the light came from stars like the Sun, and there was no dark matter, the mass-to-light ratio would be 1.) This implies that the light from such galaxies can be explained in terms of the emission from stars like those of our own Galaxy, but these stars make up only a fraction of the mass of the distant galaxy. Theorists such as Geoffrey Burbidge have argued that the most likely explanation of this is the presence of a central black hole in these elliptical galaxies, to which the visible bright stars, and perhaps some dark material, are bound. Although there is no similar mass-to-light problem for spiral galaxies, other evidence has been interpreted as suggesting that these could also contain collapsed matter, and there may even be such an object at the centre of our own Galaxy, where many energetic processes seem to operate.

Now, it's very difficult to account for the evolution of a black hole—galaxy system simply in terms of the gravitational collapse of a diffuse gas cloud, although this *is* the standard picture of galaxy formation. The only alternative is that something unusual—the central collapsed object—must have been present from the earliest stages of the evolution of the Universe. The galaxy, then, grew up around it. If the retarded core exploded later as a cosmic gusher, the outbursts from many galaxies and quasars would be neatly explained.

There have been arguments about whether or not a collapsed object could exist for a long time in the expanding Universe without attracting so much matter that it turned into a large black hole, and this is one of the difficulties that makes Narlikar liken the situation today to that of black hole theories of the 1960s. But if such objects can persist as the Universe expands, there is no doubt about the kind of galaxy that builds up around them.

The gravitational field from a retarded core in an expanding Universe will act to restrain the expansion of nearby gas, and there seems no reason to doubt that stars will form in the resulting gas cloud. Once again, the standard Einstein—de Sitter equations of cosmology can be adapted to describe the situation . . . with provocative results.

Calculations show that for a typical elliptical galaxy (with a radius of a few times 10^{22} cm) the mass of the retarded core needed to hold the system together as the Universe expands is about a billion times that of the Sun. The mass of the star cloud produced is yet a thousand times more massive still. These figures tie in exactly with the observed dimensions of large elliptical galaxies and with the calculations made by Burbidge and colleagues. If the mass-to-light problem is to be resolved by a collapsed object explanation, then the size of the core object must be just about a billion solar masses. In addition, the smoothly decreasing density of the gas cloud gathered around such a core would produce a smoothly decreasing density of stars in the resulting galaxy, and its intensity would appear to us to fall off from the centre as $r^{-5/3}$, *where* r is the distance from the nucleus. That is exactly the variation seen in elliptical galaxies, although the rough rule of thumb used is an inverse square law, which would be equivalent, of course, to $r^{-6/3}$.

A final 'coincidence' comes from looking at the energy requirements of strong radio sources in terms of the mass equivalent, that is, the amount of material that must be completely converted into energy to produce the observed activity. This is about 10 million solar masses. If we had a retarded core of 100 billion solar masses, as suggested by the simple calculations, then each outburst from an active galaxy would use up only 1 per cent of the material in the core. Our cosmic gusher, it seems, would have plenty of energy left for repeated outbursts, and that is just what we need to explain the many different kinds of energetic activity observed in the Universe.

We are led, once again, to more speculative ground. If we accept that these ideas are still at a very preliminary stage and that they will inevitably be modified as the theory of cosmic gushers develops, we can take a look at some of the speculations which outline the way such developments might occur. Some of the most curious extragalactic objects show seemingly interacting galaxies, and long before the retarded core hypothesis became widely discussed, observers such as Halton Arp speculated that in some cases we might actually be seeing pairs of galaxies in which one member had been ejected from the other—a galaxy giving birth to a new galaxy in much the same way that a cell divides. It needs little imagination to build from this idea, combining it with the concept of retarded cores, to suggest that in some cases the core might split into two portions, not necessarily of equal size, each becoming the nucleus of a galaxy. Could the material of the spiral arms of spiral galaxies mark the trail where a lump of superdense material has been ejected from a galactic nucleus and shot outwards, scattering material around it as it went? Perhaps that takes us a little too close to the borders of science fiction, although it has been suggested quite independently that the beautiful globular clusters of our own and other galaxies may be centred on collapsed objects in the same way that the retarded core idea sees elliptical galaxies as centred on collapsed objects.

Surprising as it may seem, there is already more than just speculation to consider when looking at how collapsed objects in galactic nuclei might fragment, and at what the end products would look like. In one study, José Luis Sérsic, an Argentinian astronomer, has calculated the dynamics of two or more 'inobservable entities' orbiting in the nuclear regions of a galaxy—just the situation we would have if our retarded core either fissioned or ejected discrete lumps of material. Although these calculations, made in 1972, did not have the concept of cosmic gushers specifically in mind, they can be fitted very neatly into the framework of that picture of the Universe.

Sérsic looked at what happens to a giant elliptical galaxy whose nucleus 'goes through a violent crisis, radiating mass energy and ejecting a massive body'. Not surprisingly, the appearance of the end products depends on the size of the galaxy and just how big a crisis it goes through, but some of the calculations come up with objects remarkably like some of the most puzzling phenomena seen in the Universe.

At the small-scale end of events, it is easy to produce models of normal-looking elliptical galaxies with small activity in the nuclei and a little radio emission—rather like many run-of-the-mill galaxies such as M87. The next step up in activity produces a dumb-bell-like structure, a galaxy with two nuclei, and this fits in with a whole class of known objects, the cD galaxies. But then things get really interesting. For even more violent crises, the ejection of materials is accompanied by an effect like the blowing of a smoke ring, but with the ring made up of visible, glowing galactic material. At different stages of development, a galaxy involved in this process might appear to us as a cigar-shaped ellipsoid with a 'band' around it, like a large and small galaxy moving apart with a ring between them, or eventually, after the ejected compact body has flown far away from its parent galaxy, we would see just the original galaxy with the 'smoke ring' alongside it. All of these forms of unusual galaxies and galaxy pairs have been photographed in the real Universe, and attempts to find an explanation for the origin of the peculiar ring galaxies have been taxing astronomers for some time. The explanations put forward so far, which include rather precise collisions between a galaxy and a cloud of gas in space, which might strip off a ring of material from the galaxy, sound no less like science fiction than the ideas associated with the ejection of material from a compact nucleus—and unlike the other exotic theories, the retarded core idea not only provides the bonus of an explanation for the existence of galaxies in the first place but also, through such work as that of Professor Narlikar, ties in with the cosmic gusher theory of other energetic astrophysical phenomena.

I have done no more here than sketch in the outlines of an area of astronomical research which is only now coming fully into its own and receiving the attention it deserves. By giving a brief overview of the state of the art with regard to white holes, I hope that I may have opened a window through which many non-specialists will be able to watch as the subject develops and to learn with the experts. The first steps in the theoretical development will probably come from including calculation of the effects of non-spherical distributions and of electric charge, two considerations which proved of vital importance in the development of black hole theories. The relative success of the concept, even in a clearly oversimplified form, augurs well for the prospect when these touches of realism are added. Although the calculations themselves may be far from straightforward, we will at least know what it means in physical terms when someone comes up with a claim that white hole theory can explain, say, energy production in quasars. This would confirm, at last, that these enigmatic objects are cosmic gushers, the very fountainheads which bring matter, energy and ultimately life itself into our Universe.

fourteen

Time and the Universe

▼

THE MOST IMPORTANT FEATURE of our world is that night follows day. The dark night sky shows us that the Universe at large is a cold and empty place, in which are scattered a few bright, hot objects—the stars. The brightness of day shows that we live in an unusual part of the Universe, close to one of those stars—our Sun, a source of energy which streams across space to the Earth and beyond. The simple observation that night follows day reveals some of the most fundamental aspects of the nature of the Universe and of the relationship between life and the Universe.

If the Universe had existed for an eternity and had always contained the same number of stars and galaxies as it does today, distributed in more or less the same way throughout space, it could not possibly present the appearance that we now observe. Stars pouring out their energy, in the form of light, for eternity would have filled up the space between themselves with light, and the whole sky would blaze with the brightness of the Sun. The fact that the sky is dark at night is evidence that the Universe in which we live is changing and has not always been as it is today. Stars and galaxies have *not* existed for an eternity, but have come into existence relatively recently. There has not been time for them to fill the gaps in between with light. Astrophysicists, who study the way in which the stars produce their energy by nuclear reactions deep in their hearts, can also calculate how much light a typical star can pour out into space during its lifetime. The supply of nuclear fuel is limited, and the amount of energy a star can produce, essentially by the conversion of hydrogen into helium, is also limited. Even when all the stars in all the galaxies in the known Universe have run through their life cycles and become no more than cooling embers, space—and the night sky—will still be dark. There is not enough energy available to make enough light to brighten the night sky. The oddity, the strangeness of the observation that night follows day, is not that the sky is dark, but that it should contain *any* bright stars at all. How did the universe come to contain these short-lived (by cosmological standards) beacons in the dark?

That puzzle is brought home with full force by the light of the Sun in the daytime. It represents an imbalance in the Universe, a situation in which there is a local deviation from equilibrium. It is a fundamental feature of the world that things move towards equilibrium. If an ice cube is placed in a cup of hot coffee, the liquid gets cooler and the ice melts as it warms up. Eventually, we are left with a cup of lukewarm liquid, all at the same temperature, in equilibrium. The Sun, born in a state which stores a large amount of energy in a small volume of material, is busily doing much the same thing, giving up its store of energy to warm the Universe (by a minute amount) and, eventually, cooling into a cinder in equilibrium with the cold of space. But 'eventually' for a star like the Sun means a time span of several thousand million years. During that time, life is able to exist on our planet (and presumably on countless other planets circling countless other stars) by feeding off the flow of energy out into the void.

Because night follows day, we know that there are pockets of non-equilibrium conditions in the Universe. Life depends on the existence of those pockets. We know that the Universe is changing, because it cannot always have existed in the state we observe today and still have a dark sky. The Universe as we know it was born, and it will die. And so we know, from this simple observation, that there is a direction of time, an arrow pointing the way from the cosmological past into the cosmological future.

All these features of the Universe are bound up with what British astronomer Arthur Eddington called the supreme law of nature. It is named the second law of thermodynamics and was discovered, during the nineteenth century, not by astronomical studies of the Universe but from very practical investigations of the efficiency of the machines that were so important during the Industrial Revolution—steam engines.

It may seem odd that such an exalted rule of nature should be the 'second' law of anything, but the first law of thermodynamics is simply a kind of throat-clearing, a statement that heat is a form of energy, that work and heat are interchangeable, but that the total amount of energy in a closed system is always conserved. For example, if our coffee cup is a perfect insulator, once the ice cube has been dropped into the hot coffee the total energy inside the cup stays the same. This is true even though the ice gets warm and the coffee gets cool. In itself this was an important realization to the pioneers of the Industrial Revolution, but the second law goes much further.

There are many different ways of stating the second law, but they all have to do with the features of the Universe that I have already mentioned. A star like the Sun pours out heat into the coldness of space; an ice cube placed in hot liquid melts. We never see a cup of luke-warm coffee in which an ice cube forms spontaneously while the rest of the liquid gets hotter, even though the two states, (ice cube plus hot coffee) and (luke-warm coffee), contain exactly the same amount of energy. Heat always flows from a hotter object to a cooler one, never from the cooler to the hotter. Although the amount of energy is conserved, the distribution of energy can only change in certain ways, irreversibly. Photons (particles of light) do not emerge from the depth of space to converge on the Sun in just the right way to heat it up and drive the nuclear reactions in its core in reverse.

Stated like this, it is clear that the second law of thermodynamics also defines an arrow of time and that this is the same as the arrow of time defined by the observation of the dark night sky. Another definition of the second law involves the idea of information—when things change, there is a natural tendency for them to become more disordered, less structured. There is a structure in the system (ice cube plus hot coffee) that is lost in the system (luke-warm coffee). In everyday terms, things wear out. Wind and weather crumble stone and reduce abandoned houses to piles of rubble; they never conspire to create a neat brick wall out of debris. Physicists can describe this feature of nature mathematically, using a concept called entropy, which we can best imagine as a negative measure of information, or of complexity. Less order in a system corresponds to more entropy. The second law says that in any closed system, entropy always increases (or, at best, stays the same) while complexity decreases.

The concept of entropy helps to provide the neatest and best version of the second law, but one which is really useful only to mathematical physicists. Rudolf Clausius, a German physicist who was one of the pioneers of thermodynamics, summed up the first and second laws thus in 1865: 'The energy of the world is constant; the entropy of the world is increasing.' Equally succinctly, some unknown modern wit has put it in everyday language: 'You can't get something for nothing; you can't even break even.' This is apposite because entropy, and the second law, also tell us something about the availability of useful energy in the world. Since energy is conserved, there can hardly be an energy 'crisis' in the sense that we are using up energy. When we burn oil or coal, we simply turn one (useful) form of energy into another (less useful, less concentrated) form. Along the way, we increase the entropy of the Universe and diminish the quality of the energy. What we are really faced with is not an energy crisis, but an entropy crisis.

Life, of course, seems to be an exception to this rule of increasing entropy. Living things—a tree, a jellyfish, a human being—take simple chemical elements and compounds and rearrange them into complex structures, highly ordered. But they are able to do so only by using energy that comes, ultimately, from the Sun. The Earth, let alone an individual living thing on the Earth, is not a closed system. The Sun is constantly pouring out high-grade energy into the void. Life on Earth captures some of it (even coal and oil are stored forms of solar energy, captured by living things millions of years ago), and uses it to create complexity, returning low-grade energy to the Universe. The local decrease in entropy represented by the life of a human being, a flower or an ant is more than compensated for by the vast increase in entropy represented by the Sun's activity in producing the energy on which that living thing depends. Taking the Solar System as a whole, entropy is always increasing.

The whole Universe—which must, by definition, be a closed system in this sense of the term—is in the same boat. Concentrated, 'useful' energy inside stars is being poured out and spread thinly throughout space, where it can do no good. There is a struggle between gravity and thermodynamics. Gravity pulls stars together and provides the energy that heats them inside to the point where nuclear fusion begins. Thermodynamics, seeking to smooth out the distribution of energy in accordance with the second law, results in pressure that counteracts the inward pull of gravity. This struggle is the story of the Universe: the struggle between gravity and thermodynamics. When, or if, the whole Universe is at a uniform temperature, there can be no change because there will be no net flow of heat from one place to another. There will be no order left in the universe, simply a uniform chaos in which processes like those which have produced life on Earth are impossible. Many scientists of the nineteenth century—and even later—worried about this 'heat death' of the Universe, an end implicit in the laws of thermodynamics. None seem to have appreciated fully that the corollary of the changes we see going on in the Universe is that there must have been a birth, a 'heat birth', at some finite time in the past which created the non-equilibrium conditions we see today. And all would surely have been astonished to learn that in all but the most trivial detail the 'heat death' has already occurred.

Energy at high temperatures is low in entropy and can easily be made to do useful work. Energy at low temperatures is high in entropy and cannot easily be made to do work. This is straightforward to understand, since energy flows from a hotter object to a cooler one. It is easy to find a cooler object than, say, the surface of the Sun, into which energy from the Sun can be made to flow and do work along the way. It is hard to find an object colder than, say, an ice cube, so that we can extract heat from the ice cube and use it to do work. On Earth, it is much more likely that heat will flow *into* the ice cube. Things would be a little different in space, where it is much colder than the surface of the Earth. An ice cube at 0°C would still contain some useful energy which could be extracted and made to do work under those conditions. But still, there is a limit, an absolute zero of temperature, 0 K on the Kelvin scale (named after Lord Kelvin, another of the thermodynamic pioneers). An object at 0 K contains no heat energy at all.

Space itself is not quite as cold as 0 K. The energy that fills the space between the stars is in the form of electromagnetic radiation, or photons. The energy of these photons can be described in terms of temperature—sunlight contains energetic, high-temperature photons, while the heat radiated by your body is in the form of lower-energy, cooler photons, and so on. One of the greatest discoveries of experimental science was made in the 1960s, when radio astronomers found a weak hiss of radio noise coming uniformly from all directions in space. They called it the cosmic background radiation. The hiss recorded by our radio telescopes is produced by a sea of photons with a temperature of only 3 K that is thought to fill the entire Universe.

This discovery was the single key fact that persuaded cosmologists that the Big Bang theory is a good description of the Universe in which we live. Studies of distant galaxies had already shown that the Universe today is expanding, with clusters of galaxies moving farther apart from one another as time passes. By imagining this process wound backwards in time, some theorists had argued that the Universe must have been born in a superdense, superhot state, the fireball of the Big Bang. But the suggestion did not meet with general acceptance until the discovery of the background radiation, which was quickly interpreted as the leftover radiation from the Big Bang fireball itself.

According to the now-standard view of the birth of the Universe, during the Big Bang itself the Universe was filled with very hot photons, a sea of highly energetic radiation. As the Universe has expanded this radiation has cooled, in exactly the same way that a gas cools when it is allowed to expand into a large empty volume. This is the basic process that keeps the inside of your refrigerator cool. When a gas is compressed, it gets hot—you can feel the process at work when you use a bicycle pump. When a gas expands, it cools. And the same rule applies if the 'gas' is actually a sea of photons.

During the fireball stage of the Big Bang, the sky was ablaze with light throughout the Universe, but the subsequent expansion has cooled the radiation all the way down to 3 K (the same expansion effect helps to weaken starlight, but not enough to explain the darkness of the sky if the Universe were *infinitely* old). The amount of everyday matter in the Universe is very small, and the volume of space between the stars and galaxies is very large. There are many, many more photons in the Universe than there are atoms, and almost all of the entropy of the Universe is in those cold photons of the background radiation. Because those photons are so cold, they have a very high entropy, and the addition of the relatively small number of photons escaping from stars today is not going to increase it by very much more. This is why it is true to say that the heat death of the Universe has already occurred, in going from the cosmic fireball of the Big Bang to the cold darkness of the night sky today. We live in a Universe that has very nearly reached maximum entropy already, and the low-entropy 'bubble' represented by the Sun is far from being typical.

The expansion of the Universe also provides us with an arrow of time—still pointing in the same direction—from the hot past to the cold future. But there is something rather odd about all this. An arrow of time—the change and decay that are fundamental features of the Universe at large and of everyday things that we are used to on Earth—is what we see on what physicists call a macroscopic scale. But when we look at the world of the very small, atoms and particles (what physicists call the microscopic world, although we are really talking of things far too small to see even with a microscope), there is no sign of a fundamental asymmetry of time in the laws of physics. Those laws 'work' just as well in either direction, forwards or backwards in time. How can this be reconciled with the obvious fact that time flows, and things wear out, in the macroscopic world?

In real life, things wear out, and there is an arrow of time. But according to the basic laws of physics developed by Newton and his successors, nature has no inbuilt sense of time. The equations that describe the motion of the Earth in its orbit around the Sun, for example, are time-symmetric. They work as well 'backwards' as they do 'forwards'. We can imagine sending a spaceship high above the Earth, out of the plane in which the planets orbit around the Sun, and making a film showing the planets going around the Sun, the moons going around the planets, and all these bodies rotating on their own axes. If such a film were made and were then run through a projector backwards, it would still look perfectly natural. The planets and moons would all be proceeding in the opposite direction round their orbits and spinning in the opposite sense on their axes, but there is nothing in the laws of physics which forbids that. How can this be squared with the idea of an arrow of time?

Perhaps it is better to pinpoint the puzzle by looking at something closer to home. Think of a tennis player, standing still and bouncing a tennis ball on the ground, repeatedly, with a racket. Once again, if we made a film of this activity and ran it backwards, it wouldn't look at all odd. The act of bouncing the ball is reversible, or time-symmetric. But now think of the same person lighting a bonfire. He or she might start with a neatly folded newspaper, which is spread into separate sheets which are crumpled up and piled together. Bits of wood are added to the pile, a burning match applied, and the fire takes hold. If that scene were filmed and projected backwards, everyone in the audience would immediately know that something was wrong. In the real world, we never see flames working to take smoke and gas out of the air and combine them with ash to make crumpled pieces of paper, which are then carefully smoothed by a human being and neatly folded together. The bonfire-making process is irreversible: it exhibits an asymmetry in time. So where is the difference between this and bouncing a tennis ball?

One important difference is that in the bouncing ball scenario we simply have not waited long enough to see the inevitable effects of increasing entropy. If we wait long enough, after all, the tennis player will die of old age. Long before that the tennis ball will wear out (and I am not even considering the biological needs of the tennis player involving food and drink). Even the example of the planets orbiting the Sun is not really reversible. In a very, very long time (thousands of millions of years) the orbits of the planets will change because of tidal effects. The rotation of the Earth, for example, will get slower while the Moon moves farther away from its parent planet. A physicist armed with exquisitely precise measuring instruments could detect these effects from even a relatively short segment of our film, and deduce the existence of the arrow of time. The arrow is always present in the macroscopic world.

But what of the microscopic world? In school, we are taught that the atoms that make up everyday things are like hard little balls, which bounce around and jostle one another in precise obedience to Newton's laws. Neither the laws of mechanics nor the laws of electromagnetism have an inbuilt arrow of time. Physicists like to puzzle over these phenomena by thinking about a box filled with gas, because under those conditions atoms behave most clearly like balls bouncing off one another. When two such spheres, moving in different directions, meet each other and collide, they bounce off in new directions, at new speeds, given by Newton's laws. If the direction of time is reversed, then the reversed collision also obeys Newton's laws. This raises some curious puzzles.

One of the standard ways to demonstrate the second law of thermodynamics is with the (imaginary) aid of a box divided into two halves by a partition. Imagine one half of such a box filled with gas, and the other empty (this is only a 'thought' experiment; we don't actually need to carry it out because our everyday experience tells us what will happen). When the partition is removed, the gas from the 'full' half of the box will spread out to fill the whole box. The system becomes less ordered, its temperature falls, and entropy increases. Once the gas is in this state, it never organizes itself so that all of the gas is in one half of the box once again. If it were, we could put the partition back and restore the original situation. That would involve decreasing entropy. On the macroscopic scale, we know that it is futile to stand by the box, partition in hand, and wait for an opportunity to trap all of the gas in one end.

But now look at things on the microscopic scale. The paths followed by all of the atoms of gas in moving out from one half of the box—their trajectories—all obey Newton's laws; and all the collisions the atoms are involved in along the way are, in principle, reversible. We can imagine waving a magic wand, after the box has filled uniformly with gas, and reversing the motion of every individual atom. Surely, then, they would all retrace their trajectories back from whence they came, retreating into one half of the container? How is it that a combination of perfectly reversible events on the microscopic scale has conspired to give an appearance of irreversibility on the macroscopic scale?

There is another way of looking at this. In the nineteenth century, the French physicist Henri Poincaré showed that such an 'ideal' gas, trapped in a box, from the walls of which the atoms bounce with no loss of energy, must eventually pass through every possible state that is consistent with the law of conservation of energy, the first law of thermodynamics. Every arrangement of atoms in the box must happen sooner or later. If we wait long enough, the atoms moving about at random inside the box must all end up in one end, or indeed in any other allowed state. Putting it another way, if we wait long enough, the whole system must return once again to any starting point.

'Long enough', however, is the key term here. A small box of gas might contain 10^{22} atoms (that is, 1 followed by 22 zeros), and the time it would take for them to return to any initial state would probably be many, many times the age of the Universe. Typical 'Poincaré cycle times', as they are known today, have more zeros in the numbers than there are stars in all the known galaxies of the Universe put together, numbers so big that it doesn't really make any difference whether you are counting in seconds, or hours, or years. Putting it another way, those huge numbers represent the odds against any particular state occurring, by chance, during any particular second, hour, or year that you happen to be watching the box of gas.

This provides the standard 'answer' to the puzzle of how a world that is reversible on the microscopic scale can be irreversible on the macroscopic scale. The irreversibility, the traditionalists allege, is an illusion. The law of increasing entropy is a statistical law, they say, in the sense that a decrease in entropy is not so much specifically forbidden as extremely unlikely. If we watch a cup of luke-warm coffee for long enough, according to this interpretation, it will indeed spontaneously produce an ice cube while the surrounding liquid gets hotter. It just happens that the time required for this to occur is so much longer than the age of the Universe that we can, for all practical purposes, ignore the possibility. (If this sounds vaguely familiar, perhaps you've been reading *The Hitch Hiker's Guide to the Gal-*

axy, in which Douglas Adams describes the workings of the 'infinite improbability drive'.) But this interpretation of the law of increasing entropy as a statistical rule, not an absolute law of nature, has recently been questioned.

New ideas in thermodynamics turn the traditional interpretation on its head, and hold that the irreversibility really is a fundamental feature of our Universe, once the arrow of time is defined. They stem mainly from the work of Ilya Prigogine, who was born in Moscow in 1917 but has been associated with the Free University of Belgium since 1947, and subsequently also with the University of Texas, Austin. He received the Nobel Prize for Chemistry in 1977 for his work on non-equilibrium thermodynamics. But his ideas have yet to filter through into the textbooks used by most students of thermodynamics, even at university level.

Prigogine's attack on the problem of reconciling macroscopic irreversibility and microscopic reversibility can be understood in terms of the Poincaré recurrence time, by taking on board some basic ideas from quantum theory. Quantum physics, developed in the first half of the twentieth century, provides a better description of the behaviour of atoms and smaller particles than the older, classical ideas of electromagnetism and Newtonian mechanics. It is only with the aid of quantum physics that modern scientists are able to understand the workings of atoms, and the interaction between particles and electromagnetic fields. We don't need to go into all the details here, but there are two features of quantum physics that are relevant to the thermodynamics of the Universe.

The first important point is that the equations of quantum physics, like those of classical physics, are time-symmetric. There is no arrow of time built in to quantum physics, and reactions, or interactions, can, according to the equations, proceed just as happily 'backwards' as 'forwards'. That seems to leave us in the same bind. But the second salient feature of the new physics gets us off that particular hook.

Werner Heisenberg, a German scientist who made major contributions to the development of quantum theory, discovered that the equations do not allow us to make a precise measurement of both the position and the momentum of a particle at the same time. We cannot know, as a matter of principle, exactly where a particle is and where it is going. We can determine either property on its own as accurately as we like, but the more precisely we measure position the less information we have about momentum, and vice versa.

When Heisenberg first reported this uncertainty principle, many people thought that he was talking about some limitation on the practical skills of human observers, and meant that although an electron, say, could be in a definite position and be moving with a definite velocity and momentum, it was forever beyond our skill to measure them both at the same time. But this is wrong. The essential feature of Heisenberg's discovery is that the entity we call 'an electron' does not possess both a well-defined position and a well-defined momentum simultaneously. There is an *intrinsic* uncertainty, which has to do with the way the Universe is put together and has nothing to do with the skill, or otherwise, of human experimental physicists.

This is not common sense, but why should it be? Our common sense is based on everyday experience with objects on the human scale, and on that scale the uncertainty effect is far too small to notice. We have no basis for knowing what the 'common sense' of things on the scale of atoms and electrons is, except with the aid of theories that predict how collections of such particles will respond in certain circumstances. The theory that makes the best, most accurate and most consistently correct predictions is quantum theory, including the uncertainty principle. Indeed, this is only the tip of the iceberg of quantum oddity, for the best interpretation of the 'meaning' of quantum theory is that there is no underlying 'reality' which builds up to make the macroscopic world. The only reality lies in the actual events we observe—the swing of a needle across a dial when an electric current flows, the click of a Geiger counter as a charged particle passes through its detector, and so on. Nothing is real unless it is observed, say the quantum physicists, and there is no point in trying to imagine what atoms and electrons are 'doing' when they are not being monitored.

All of these ideas carry over into Prigogine's version of thermodynamics. The reality that we observe is the macroscopic world, with its inbuilt arrow of time and asymmetry. Why, he asks, should we imagine that this world is built up in some way from the behaviour of countless tiny particles obeying precisely reversible, time-symmetric laws of behaviour? To Prigogine, the macroscopically derived second law of thermodynamics is the fundamental truth, a precise law that always holds, not a statistical rule of thumb that applies, more or less, for most of the time. It is the apparently time-symmetric behaviour of little spheres bouncing off one another that he regards as an approximation to reality. 'Irreversibility,' he says, 'is either true on all levels or none. It cannot emerge as if by a miracle, by going from one level to another.'

We can see what he is driving at, and the direct relevance of quantum physics to thermodynamics, by looking at another example of a closed system, the kind which Poincaré said ought to return to its initial conditions, given enough time. We start, once again, with a box full of gas, but this time make it a little more complicated by placing in the box a smoothly sloping hill of material, completely symmetrical, rising up to a rounded top. Imagine a perfectly round ball balanced precisely on top of that hill, with the box shut, as usual, to keep it thermodynamically isolated from the rest of the Universe. What will happen to the ball? Obviously, it will roll off the top of the hill. But which way will it roll? The direction taken by the ball, and the subsequent history of the material in the box, will depend on some tiny accumulation of little nudges by the atoms of gas bouncing off it. There will be a minuscule pressure pushing the ball one way, just by chance, and off it will roll.

According to Poincaré, the ball will eventually return to its starting place. When the ball rolls off the hill, it gives up energy to the gas, energy derived from the fall of the ball, ultimately from gravity. If we wait long enough (many, many times the age of the Universe!) it will just happen that most of the atoms of gas bouncing off the ball will be moving in the same direction and will give it a little push, with precisely the same amount of energy that it previously gave up to the gas, so that it rolls back up the hill while the gas cools down. There will be other occasions when the ball receives a push in the wrong direction, or one that is too strong, or too weak, to leave it balanced once again on top of the hill. But after a suitably long interval there will be an occasion when the push is exactly sufficient to return the ball to the top of the hill, and leave it balanced there. The system has returned, as predicted, to its original state. Or has it?

If there is the tiniest difference in the way the atoms of gas now strike the ball, compared with the first time it was on top of the hill, it will roll off in a different direction, and the future history of the little world inside the box will be completely different. And there must be tiny differences in the way the atoms are striking the ball, because quantum uncertainty makes it impossible to define any set of conditions precisely for the atoms. Even in this very simple case, we can imagine the ball so precisely balanced that as tiny a change in the conditions as you like will alter its future behaviour. The real Universe is vastly more complicated than this, and complex systems involving many particles are known to be prone to very strong instabilities, so that a tiny change in starting conditions produces a drastic alteration in the system's behaviour.

Or, if you prefer, think of things in terms of the actual reversibility of atoms moving in a box of gas. When we think about the system in which the gas in half of the box spreads out to fill the whole box, it is easy to say 'imagine reversing the motion of every atom simultaneously.' This conjures up an image of something like a pool table, with balls moving on it, which suddenly reverse their trajectories and return to their starting positions. We can't actually do the trick, but we can, indeed, imagine it. But think what the simple statement really means. It requires that the position of every atom should be precisely determined, and that its velocity should simultaneously be precisely determined and then exactly reversed while the atom stays in precisely the same position. But quantum physics tells us that this is impossible! No atom has the two characteristics of precise position and precise velocity at the same time! The laws of nature, as they are best understood today, make it impossible to reverse the direction of every atom in the gas, *in principle*, not just because of the practical limitations set by human skills. There is no magic powerful enough to do the job. So, once again, we find that a system that seems to be reversible is, in fact, irreversible.

This is just a very simple interpretation of one aspect of Prigogine's reinterpretation of thermodynamics. But the gist of the message is plain, and important. No matter how long we sit and watch a luke-warm cup of coffee, it will never spontaneously give birth to an ice cube and heat up; no matter how long we sit by a box of gas, it will never all congregate in one half of the box, so that we can trap it in a state of lower entropy. The second law of thermodynamics is an absolute ruler of the Universe. We live in an irreversible Universe, with a well-defined arrow of time. And that is why the sky is dark at night.

fifteen

Weighing Up Empty Space

▼

THE IDEA THAT EMPTY space can have weight might appear fanciful, or even meaningless: how can 'nothing' weigh something? But empty space is far from being nothing. Even when all particles are removed from a region of space it will still contain 'virtual' particles, temporarily created from quantum processes. They bestow upon the vacuum both energy and pressure. The famous uncertainty principle of quantum mechanics actually allows pairs of particles, such as a positron and an electron, to pop briefly into existence out of nothing at all, provided they annihilate one another before the Universe has time to 'notice'. Photons (particles of light) are also allowed to flicker in and out of existence in this way, contributing their share of energy (and therefore mass, in line with Einstein's relation $E = mc^2$) to the vacuum, and one might expect this vacuum mass to gravitate.

Unfortunately, finding out how much empty space weighs does not simply involve putting an empty box on a set of scales and weighing it. Space is all around us, and if it does gravitate, it will pull equally in all directions. The only way this gravitational pull will manifest itself is in the motion of the Universe as a whole. Attempts have been made to detect any effect of this kind in the Universe today. If the weight of space remained, by even the most minute amount, different from zero, this would show up in the way the Universe expands, by competing with the way that the attractive force of ordinary gravitating matter gradually slows the expansion.

No such effect has been detected, and it is clear that any weight of space must be very small indeed. But this conclusion presents something of a mystery. Quantum theory does not require the energy of the vacuum to be small. In fact, calculations suggest that it should have a truly enormous value, about 120 powers of ten greater than the maximum value permitted by the observations! The number 10^{-120}, which represents the ratio of the maximum actual vacuum energy to the calculated value, is so small that it is tempting to believe that the weight of space is precisely zero. So we have the curious situation that a colossal vacuum energy seems to be the 'natural' state of affairs, while the observed value of zero, or close to it, seems peculiar—even contrived.

Why contrived? The accuracy of the term becomes clear when we try to understand how the present value could be so small. Quantum vacuum energy can, in fact, be either positive or negative. If nature could arrange for the negative and positive contributions to cancel out, the result would be zero. But that would require a Cosmic Accountant with an exquisite skill at balancing the books. In order to get rid of this unwelcome contribution, the influences of different kinds of particles and fields have to be delicately arranged to cancel each other out to astonishing precision. It seems highly implausible that this would occur accidentally. The only alternative is for there to be a natural mechanism that *forces* the weight of space to be zero. To find such a mechanism, we have to peer into the mysteries of black holes.

Most people have an instinctive fear of vast spaces. It is an antipathy evidently shared with our ancestors who, appalled at the prospect of an infinite void, preferred to believe that the Universe was confined by concentric shells. Even the idea of space between atoms makes some people uncomfortable. Many Greek philosophers reacted strongly to the assertion by the Atomists that the world consists of particles moving in a void, an antipathy captured by the tag 'Nature abhors a vacuum'. It seems that the concept of empty space touches on some deep atavistic fears buried in the human psyche. No wonder, then, that people have an awe-filled fascination for the relatively recent speculation that they could be swallowed up by empty space.

One of the best-selling science books of all time was John Taylor's *Black Holes*, published in 1973. Although the idea of black holes in space had been shaping up among scientists for some time, they had only been given that evocative name in the late 1960s, and it was not until the 1970s that the general public became aware of them. The bizarre and frightening properties of these objects guaranteed immediate attention and ensured that the term 'black hole' found a permanent place in the English language. These days, it is almost commonplace to read about black holes lurking at the centres of galaxies, busily eating away at the Universe. But only thirty years ago, they remained a wild speculation.

Black holes form when gravity, the weakest of nature's four forces, rises to overwhelming proportions. The power of gravity to grow without limit is due to its universally attractive nature and its long range. The other forces are all self-limiting. The nuclear forces are confined to a subatomic range, while electromagnetic forces come in both attractive and repulsive varieties, which tend to cancel out. But go on adding more and more matter to an object, and its gravity will continue to rise without limit.

The gravity at the surface of an object depends not only on its total mass, but also on its size. For example, if the Earth were compressed to half its present radius, we would all weigh four times as much. This is because the force of gravity obeys an inverse square law—it gets stronger at shorter distances. The higher gravity would make it harder to escape from a shrunken Earth. At its present size, an object has to be propelled upwards from the surface of our planet at about 11 kilometres a second—the so-called escape velocity—before it will fly off into space for good. For a half-size Earth, the escape velocity is about 41 per cent greater.

If the Earth could be progressively shrunk (retaining the same mass), its surface gravity and the escape velocity would go on rising. When the compressed Earth was reduced to the size of a pea, the escape velocity would reach the speed of light. This is the critical size. It implies that no light could escape from such an object, and so the shrunken Earth would effectively disappear from view. It would become, as far as anyone outside was concerned, completely black.

There is no risk that the Earth will shrink in the manner described, for its gravity is safely countered by the solidity of its material. But in the case of larger astronomical objects the situation is different. Stars like the Sun are engaged in a ceaseless battle with gravity. These balls of gas are prevented from collapsing under their own weight by a huge internal pressure. The core of a star has a temperature of many millions of degrees, and this heat produces a pressure sufficient to support the colossal weight of the overlying layers of gas. But this state of affairs cannot last forever. The internal heat is generated by nuclear reactions, and eventually the star runs out of nuclear fuel. Ultimately, the pressure support must fail, and the star will be left to the mercy of gravity.

What happens next depends crucially on the total mass of the star. A star like the Sun will end its days by shrinking to about the size of the Earth, and it will then become what astronomers call a white dwarf. Such stars have long been known to exist; for example, the bright star Sirius has a white dwarf companion in orbit around it. Because of the compression, the surface gravity of a white dwarf is immense. A teaspoonful of the highly compressed matter there would contain about as much matter as a heavy truck does on Earth, but it would weigh 10 million tonnes in the star's strong gravitational field. White dwarfs avoid further compression with the aid of quantum mechanical effects. Their electrons resist being squeezed any closer together for reasons similar to those which explain how electrons in atoms are confined to definite energy levels, so preventing normal atoms from collapsing. It is a dramatic demonstration of quantum physics at work.

The ability of quantum effects to stabilize a star was already appreciated by the early 1930s. At that time a young Indian student by the name of Subrahmanyan Chandrasekhar was travelling to England to work with the famous Cambridge astronomer Arthur Eddington. During the long journey by boat he did some calculations and discovered that if a star had about 50 per cent more mass than the Sun, then the quantum pressure support provided by the electrons would fail, and the star would collapse further. He showed the calculation to Eddington, who refused to believe it! But Chandrasekhar was right, and heavy stars cannot become white dwarfs.

The further compression of a star, which would occur if it had enough mass for gravity to overcome the resistance of electrons, involves a transformation of the very atomic nuclei that comprise most of its mass. The crushed atoms undergo a sort of reversed beta decay, with electrons and protons being squeezed together to form neutrons. At this density the neutrons can deploy the same quantum effects as do the electrons in a white dwarf. So long as the star is not too massive, it shrinks to become a ball of neutrons packed tightly together, like a monstrous atomic nucleus. As a result of the enormous compression, the entire star may be no larger than a typical city, yet contain more material than the Sun.

These are the 'neutron stars' that have now been detected in the form of pulsars. Their surface gravity is so great that the escape velocity is an appreciable fraction of the speed of light. So we know by direct observation, from the fact that neutron stars exist, that there are objects in the universe close to the 'black star' limit.

What about stars so massive that even the neutron support fails? Astronomers are not certain of the exact limit beyond which further collapse must occur, and there is even a conjecture that a still more compact stable phase of matter—a sort of quark soup—might be possible. But a quite general limit on the mass of a collapsed star can be inferred from the theory of relativity. To support a star of a certain mass the material of its core has to have a certain rigidity. The heavier the star is, the stiffer the core material must be. But the stiffness of a substance is related to the speed with which sound moves through that substance: the stiffer the material, the faster the speed of sound. If a static collapsed object were as massive as about three Suns, it would have to be so stiff that the speed of sound would exceed the speed of light. As the theory of relativity forbids any physical influence to propagate faster than light, this state of affairs is impossible. The only route left open for the star is *total* gravitational collapse.

If an object implodes from neutron star densities, it will disappear in less than a millisecond, so intense is the pull of its gravity. The surface of the star rapidly crosses the critical radius that prevents light from escaping, and so a distant observer would no longer be able to see the object. We now know, from relativity theory, that no force in the Universe can prevent the star from continuing to collapse once it has reached the light-trapping stage. So the star simply shrinks away, essentially to nothing, leaving behind empty space—a hole where the star once was. But the hole retains the gravitational imprint of the erstwhile star, in the form of intense space and time warps. Thus the region of gravitational collapse appears both black and empty—a black hole.

By definition, we can never see inside a black hole, but theory can be used to infer what it would be like for an observer to enter a hole and explore its interior. The key to understanding the physics of black holes is the so-called event horizon. Roughly speaking, this is the surface of the hole. Any event occurring within the hole (inside the event horizon) can never be witnessed from the outside because no light (or other form of signal) can escape to convey information about such events to the outside world.

If you should find yourself inside a black hole's event horizon, not only could you never escape but, like the star that preceded you, you would be unable to halt your inward plunge. Just what happens when you arrive at the centre of the hole nobody knows for sure. According to the General Theory of Relativity, there is a so-called 'singularity' there, a boundary of space and time at which the original star (and any subsequent infalling matter) is compressed to an infinite density, and all the laws of physics break down. It may be that quantum effects cause spacetime to become fuzzy very close to the centre, with the singularity smeared out over the Planck scale, about 10^{-35} of a metre. At this stage we do not have a reliable enough theory to know, and it is no good trying to look or sending an automatic robot probe to look for you. The already fierce gravity of the hole rises without any known limit as the centre is approached. This has two effects. If you fell into the hole feet first, your feet would be closer to the centre than your head, and would be pulled progressively harder than your head, stretching your body lengthwise. In addition, all parts of your body would be pulled towards the centre of the hole, and so you would be squeezed sideways. At the end of this 'spaghettification' you would be crushed into non-existence (or lost in a haze of quantum uncertainty). All this would happen in the fraction of a second prior to your reaching the singularity. You couldn't observe it without being irreversibly incorporated into it.

If, then, an object that falls into a black hole cannot re-emerge into the outside Universe, what happens to it? As we have explained, any object that encounters the singularity is annihilated: it ceases to exist. A precisely spherical ball of ordinary matter, for example, collapsing to become a black hole, will shrink to the common centre. All the matter will be squeezed into a singularity. But what if the collapsing object is not precisely spherical? All known astronomical objects rotate to a greater or lesser extent, and as an object shrinks its rotation rate rises. It seems inevitable that a collapsing star will be spinning rapidly, causing it to bulge out at the equator. This distortion will not prevent a singularity forming, but it could be that some of the infalling material of the star will miss it.

Idealized mathematical models of charged and rotating holes have been examined to find out where the singularity lies and where the infalling material goes. The models indicate that the hole acts as a sort of bridge, or spacetime tunnel, connecting our Universe with another spacetime that is otherwise completely inaccessible to us. This astonishing result opens up the prospect of an intrepid space traveller passing through the black hole unscathed and entering another universe, arriving somewhere beyond our infinite future. If this could be accomplished, it would not be possible for the traveller to retrace the passage back to the starting point. Diving into the tunnel from the other universe would bring our daring spacefarer not back to our Universe, but to a third universe, and so on *ad infinitum*. A rotating black hole is connected to an infinite sequence of universes, each representing a complete spacetime of possibly infinite extent, and all connected together through the interior of the black hole. The prospect of making any practical sense of this is so daunting that we shall leave it to the science fiction writers (Carl Sagan used it in his novel *Contact*).

What would the remote end of the black hole bridge look like to an observer located in the other universe? According to the simplest mathematical models, an observer would see the object as a source of outward rushing material—the explosive creation of matter—often called a 'white hole'. Our universe abounds with exploding objects, such as quasars, and this has led to the conjecture that there are indeed spacetime tunnels channelling matter into our space and time through black holes from other universes. Few astrophysicists take the scenario seriously, however. In particular, they point out that the mathematical models neglect the effect of surrounding material and radiation, which would be sucked back into the white hole by gravity, turning it into a black hole. The simple models also neglect the effects of subatomic physics. More elaborate models that incorporate these features indicate that the interior of the hole is so disrupted by these disturbances that the spacetime tunnel would be smashed, and the bridges connecting our Universe with other hypothetical spacetimes are thereby blocked. The balance of opinion among the experts is that *all* matter which enters a black hole eventually encounters a singularity of some sort.

But what if quantum effects remove the singularity somehow? Unfortunately, since we have no complete quantum theory of gravity we cannot model the consequences of quantum effects in smearing out the singularity with any confidence. Whether they actually result in the complete removal of the singularity is uncertain. Some physicists expect this to be the case, and even argue that the very concepts of space and time will cease to apply under these extreme conditions. Just what sort of structures might replace them is a matter for conjecture. The safest position, therefore, is to regard the singularity as merely a breakdown of *known* physics, rather than the end of all physics.

It is easy to see how tunnels associated with black holes, if they do exist, could change the structure of curved spacetime. We represent spacetime by a two-dimensional sheet, like the surface of a piece of paper. Bending the sheet over makes a flattened U shape, with the top and bottom of the sheet brought close together, and separated by a small gap across the third dimension. If we could join the two opposite parts of the sheet by a tube through the third dimension, it would be possible to travel from one to the other through the tunnel, without going the long way around. Such a connection between different parts of the same spacetime is officially known to relativists as a 'wormhole'. Anything that we can picture happening to a two-dimensional sheet folded through a third dimension can be translated mathematically into four-dimensional spacetime folded through higher dimensions. If the two ends of the wormhole are, say, one light year apart across the 'main sheet', no signal can travel between them in less than a year through that route. But by travelling through the wormhole, a signal, or perhaps even a person, can get from one end to the other in much less than a year.

Now imagine the curved spacetime being unbent and laid flat again, with the wormhole stretching and remaining intact. This leaves you with a flat spacetime in which two different regions are connected by a U-shaped wormhole, rather like the handle on a teacup. At first sight, this is much less interesting. It looks as if the distance from one end of the wormhole to the other is longer if you travel through the wormhole than if you travel through ordinary space. This is not necessarily the case, however, because space and time behave differently inside the wormhole; even though the spacetime of the parent universe is flat (or nearly so) and the wormhole is curved, it can still act as a short-cut, so that a traveller entering one mouth of the wormhole emerges from the other mouth almost instantaneously, no matter how far away the other mouth is across the universe.

In fact, the topic of wormholes is currently being intensively investigated by many different research groups, but *not* to test the possibility of travelling through short-cuts in spacetime. Interest focuses instead on the properties of microscopic virtual wormholes. Just as quantum fluctuations in the vacuum create temporary photons, so, on an even smaller scale, they should spontaneously create temporary (virtual) wormholes. The size of these wormholes is typically twenty powers of ten (10^{-20}) smaller than an atomic nucleus. Thus, on an ultramicroscopic scale, space would be a labyrinth of such structures, endowing it with a complicated topology that has been dubbed 'spacetime foam'. With masterly understatement, such tiny tunnels through spacetime are simply referred to by relativists as 'microscopic' wormholes.

What is now being taken seriously is the possibility that virtual quantum wormholes might offer a clue to one of the great outstanding mysteries of modern physics—the lack of weight of empty space. One of the fields that contributes to the quantum vacuum energy is the gravitational field, and it is quantum fluctuations of the gravitational field that create not only baby wormholes but other distortions in the geometry of spacetime. Some of these distortions will take on the form of entire 'baby universes', connected to our own spacetime by a wormhole, as if by an umbilical cord. The whole process takes place on an ultramicroscopic scale, and one must imagine these tiny protuberances continuously fluctuating, sometimes disconnecting themselves entirely from our Universe as their umbilical wormholes pinch off, sometimes being re-absorbed back into our spacetime as the quantum fluctuation fades away. The cumulative effect is to clothe the space of our Universe with a sort of gas of shifting minispace bubbles. But each minispace is its *own* space and time. The only connection with our spacetime is through the umbilical wormholes, and the diameters of the wormhole mouths are, remember, far smaller than the diameter of an atomic nucleus, so we cannot observe them directly.

How will all this affect the nature of the vacuum? The task of computing the effect of this monstrously labyrinthine spacetime froth on the weight of the space to which it clings has been taken up by Stephen Hawking of Cambridge and Sidney Coleman of Harvard. Their calculations appeal to a universal principle of physics, known as the principle of least action. It states, roughly, that whenever anything changes, it does so in such a way as to minimize the effort. For example, a pool ball always chooses to travel along a straight path between two points rather than exert itself by following a zigzag path, unless it is acted upon by external forces. This law of natural indolence, when applied to wormhole fluctuations, implies that those baby universes with very little vacuum energy are preferred to those with a lot. The most preferred are those with precisely zero energy, so the appropriate quantum 'average' or expected value of the vacuum energy will be very close to zero, and this value leaks in to our Universe from the ever-changing myriad of baby universes to which we are connected.

If these calculations hold up, we will have arrived at a curious conclusion. Our naïve expectation that empty space is weightless turns out to be correct, but not for the reasons we thought. It has nothing to do with its emptiness as such, because even empty space is alive with quantum activity. The weightlessness is due instead to an unseen froth of parasite universes that cling to our spacetime through a network of invisible wormholes. Without wormholes, the Universe would indeed be so heavy that it would collapse.

sixteen

Particle Physics, Cosmology and the First Hundredth of A Second

▼

HAVE YOU ANY IDEA how small an atom is? In 1811, the Italian chemist and physicist Amedeo Avogadro derived his famous hypothesis: for any fixed temperature and pressure, equal volumes of gas contain the same number of molecules. Later experiments proved Avogadro's hypothesis to be correct, and established that each litre of gas—*any* gas—at a pressure of one atmosphere and a temperature of 0°C contains, in round terms, 27,000 billion billion (27×10^{21}) molecules. Clearly, atoms are very small. But how small? Converting from volumes into weights, with which most of us are more familiar, Avogadro's hypothesis can be used to tell us how many atoms there are in a gram of a pure substance. Hydrogen is the simplest atom, and Avogadro's number is the number of atoms of hydrogen in one gram of the gas. It checks in at 6×10^{23} (a 6 followed by 23 zeros), and that is a lot of atoms for a tiny puff of gas.

It would also, of course, be the number of atoms in one gram of liquid or solid hydrogen, were you able to come by such a thing. But solid lumps of hydrogen are few and far between, so in order to get a grasp of what Avogadro's number tells us about the size of atoms, it is better to think in terms of some familiar solid, like carbon. An atom of carbon weighs just about 12 times as much as an atom of hydrogen, so the same number of atoms weigh 12 times as much. *Twelve* grams of carbon contain Avogadro's number of atoms. Twelve grams (just under half an ounce) is very much the sort of weight with which we are familiar: the weight of a spoonful of sugar, a rather large diamond, or a rather small lump of coal. And that small lump of coal contains 6×10^{23} atoms. How can we put that number in perspective?

Huge numbers are often called 'astronomical', and readers of the *Griffith Observer* have come across some genuinely astronomical numbers from time to time, so let's find one which matches up to Avogadro's number, if we can. The age of the Universe might be a good place to start. Few astronomers nowadays would argue much with an age of 15 billion, or 15×10^9 years. But clearly 10^{23} is still a great deal bigger than 10^9. Let's turn the age of the Universe into an even bigger number, using the smallest unit of time with which we are comfortable, the second. Each year contains 365 days, each day 24 hours, each hour 3,600 seconds. Multiplying up, in round terms we find that each year contains 32 million, or about 3×10^7 seconds. Following the rule that in multiplying such large numbers you add the exponents (the powers of ten), we can get the age of the Universe in seconds. It is roughly 5×10^{17} seconds.

That's a whole lot bigger than 10^9, but it is still far short of 10^{23}, six powers of ten short. If we divide 6×10^{23} by 5×10^{17}, we have to subtract the exponents, and end up, near enough, with 10^6—a million. Imagine a supernatural being watching our Universe develop from the Big Bang of 'creation'. This being is equipped with a small lump of carbon, weighing just 12 grams and containing Avogadro's number of atoms. The being also has a pair of tweezers so fine that it can remove one atom from the lump of carbon every second and discard it. Starting with the first moment of the Big Bang and continuing busily until the present day, the being has been doing just that. The number of atoms removed is now 5×10^{17}, and that represents *one-millionth* of the atoms in the lump of carbon to start with. After all that activity, working steadily for 15 billion years and throwing away one atom every second, the supernatural being is still left with a million times more atoms than have been discarded. *Now* do you have some idea how small an atom is?

It is hardly suprising, now that we have some idea of the size of an atom, that the laws of physics which describe the behaviour of atoms and subatomic particles, and enable us to predict the outcome of experiments involving atoms and smaller particles, are not the same laws of physics that apply in the everyday world. The miracle is that we understand anything at all about atoms, and that the quantum theory, developed in its full form in the 1920s, really does enable us to explain how the subatomic world works. It is even more miraculous that the laws of physics that have been found to apply to the world of the very small have now also been applied to the early stages of the evolution of the Universe itself.

Because we see the Universe about us to be expanding, with distant galaxies moving farther apart from one another, it seems a logical conclusion that long ago all of the matter in the Universe was packed tightly together. This is the basis of the Big Bang theory, the root of the idea that there is indeed a definite 'age of the Universe'. That theory starts out from the observations, pioneered by Edwin Hubble and Milton Humason in the 1920s and 1930s, that the light from distant galaxies shows a redshift which is proportional to the distances of those galaxies from us. This shift of features in the spectra of light from distant objects is interpreted as an effect of recession, rather like the Doppler effect which deepens the note of the siren of a police car that is rushing away from you.

Over the past fifty years or so, astronomers have piled up a wealth of evidence that the Universe is indeed expanding and that the redshifts are indeed due to the expansion of the whole Universe. The Universe is expanding and the density of matter in it is getting less as that matter is spread more thinly across an ever-growing cosmos. In that case, if we imagine looking back in time to earlier epochs of the Universe, then galaxies must have been closer together, and the matter density of material in the Universe must have been greater. Look back far enough in time and you can imagine an epoch when all the matter in the Universe was crushed together in one seething maelstrom, a fireball of energy and particles as dense as the matter at the heart of an atom, in its nucleus.

The laws of physics that are needed to describe what happens to the particles and radiation in such a fireball are the laws of physics that apply to nuclear matter—the laws of quantum theory and particle physics. It is only a few years since cosmologists proudly told the world how those fundamental equations of physics could explain the behaviour of the fireball during the first three minutes of creation, the three minutes after the time when, by winding the present expansion of the Universe backwards, we would arrive at (or start from, depending on your point of view) a state of literally infinite density, with all of the matter of the Universe compressed into a mathematical point, a singularity. What those stories of the first three minutes tended to gloss over, however, was the fact that the theories of the early 1970s 'only' *started out* one-hundredth of a second after the outburst from the state of infinite density, the singularity itself. They started with all the matter in the visible Universe we know today squeezed into a hot fireball about as big as a basketball, and explained the evolution of the Universe from then on. That didn't seem too bad at the time. If our theories could explain 15 billion years of cosmic history, what did the odd hundredth of a second matter? But still an unanswered question remained at the back of their minds—*how* did the Universe get to be a superdense fireball the size of a basketball in the first place?

Inevitably, some theorists tried to probe further back towards the moment of 'creation'. As a result, cosmologists today speak glibly of events that occurred when the age of the Universe was 10^{-35} of a second or less, quite literally a split second after the moment of 'creation' itself. And it turns out that events in that epoch before the first hundredth of a second are every bit as interesting as what came after, and may explain how and why the Big Bang that set the Universe expanding actually happened. Not only do the laws of particle physics, deduced from atom-smashing experiments carried out with particle beams at high-energy accelerators, provide a self-consistent story of how the Universe could have got to be the way it is starting from a state of superheat and superdensity just 10^{-35} of a second after it started rolling. According to the latest ideas, the laws of particle physics may have been *responsible* for blowing the Universe up to its present size and thereby reducing its density to the state of emptiness we see all about us today. The theory is called 'inflation'.

Particle physicists find out how particles behave at high energies and densities, and they test their theories, literally by bouncing particles smaller than atoms off one another at high speeds and by 'looking' (with the aid of some sophisticated detectors) at the bits and pieces produced by the collisions. Their best modern theories explain everything about such particle interactions pretty satisfactorily, and they also make predictions about how particles should behave at still higher pressures and densities. There is no practicable hope of achieving such conditions on Earth, but if the Big Bang really did start out from a point of infinite density and pressure, then those best modern theories, thoroughly tested as far as they can be on Earth, ought to describe the behaviour of matter in the Universe once things had cooled down a bit from the infinite state.

The key concept is *symmetry*. The fundamental equations of physics are, for example, time-symmetric: they work equally well backwards as they do forwards. Other symmetries can be understood in geometrical terms. A rotating sphere, say, can be reflected in a mirror. Looking down on the top of the sphere, we may see it rotating counterclockwise. If we do, then the mirror image will seem to be rotating clockwise. Rotation is not symmetric under the transformation called reflection. There are many other kinds of symmetry in nature. The electron and positron, for example, can be thought of as mirror image particles, with a reversed negative charge equivalent to a positive charge. The two particles are in some sense asymmetric 'mirror images'. But in the same sense, the mirror image of a neutron is another neutron—it is its own 'antiparticle', and the neutron can, in that sense, be regarded as symmetric. Other symmetries are, however, much harder to grasp intuitively.

Think of a ball balanced on one step of a staircase. It has a certain amount of energy, compared with what it would have at the foot of the staircase, because it is raised up in the Earth's gravitational field. Now move the ball to another step. We have changed the potential energy of the ball by a definite amount. It doesn't matter *how* the ball got to the different step—it might have been in a rocket on a round trip to Mars. And it doesn't matter from where we measure the potential energy: the basement, or the attic, or anywhere else will do. The *difference* in potential energy between the two states is still the same. This, too, is a kind of symmetry, because the change is always the same no matter how you measure it. Because you can 'regauge' the baseline from which the energy is measured, physicists call this a gauge symmetry. And such a symmetry turns out to hold the key to the existence of the Universe as we know it.

The same sort of thing happens with electric forces. In the nineteenth century, the pioneering Scottish physicist James Clerk Maxwell investigated the phenomena of electricity and magnetism, and he found that they could be explained by one set of equations. A moving electric charge—a current—for example, generates a magnetic field, and Maxwell set up the equations which described how such fields behaved. The equations are as important to an understanding of charge and magnetism as Newton's equations were to an understanding of the behaviour of uncharged objects. And Maxwell's equations of electromagnetism are gauge-invariant, as are the equations of the equivalent theory in quantum physics, called quantum electrodynamics, or QED.

This theory of electrodynamics is so successful that physicists have used it as the model for constructing their theories to describe the behaviour of other forces which affect particles in the atom. There is the strong nuclear force, which holds particles like neutrons and protons together in the nucleus at the heart of the atom. And there is the weak nuclear force, responsible among other things for radioactive decay. The electric force is carried by a particle, called the photon. When two charged particles interact with one another, the interaction happens because photons are swapped between the two particles. The photons carry the electromagnetic force. So physicists guessed that the weak nuclear force must also be carried by a particle, which they dubbed the weak boson, even though nobody had ever detected such a beast. In fact, they decided there must be at least three such particles. Two of them, the W© and the W^-, are necessary to carry charge, which is transferred during processes like radioactive decay, and a third, called the Z, is necessary because charge is not always transferred in weak interactions. The Z is the weak-force counterpart of the photon; the W particles are, in effect, charged photons.

The mathematical symmetries involving the weak interaction and the W and Z particles were first partially worked out by Sheldon Glashow, of Harvard University, in 1960, and published in 1961. The equations turned out to require the particles to have mass, and predicted what that mass ought to be. In 1983, a team of physicists working at CERN, the European particle accelerator lab near Geneva in Switzerland, was able to measure the masses of these particles by an indirect method, and found that they were exactly in line with the prediction. This, the crowning achievement of particle physics to date, led to the award of the 1984 Nobel Prize for Physics, and gave the physicists faith in the other predictions by the kind of theory that forecast the existence of the W and Z particles. That's just as well, since some of those predictions are strange indeed.

Glashow's work was developed further by Abdus Salam in London, Steven Weinberg at Harvard and the Dutch physicist Gerard t'Hooft. They combined the theory of electromagnetism and the theory of the weak interaction into one mathematical description, as the 'electroweak' theory. Maxwell had combined electricity and magnetism into one. Now physicists were able to go a stage further. Their ultimate aim is to combine *all* the forces of nature into one mathematical description, one set of equations: a unified field theory. The electroweak theory, a step on the road to full unification, says that for very high energies there is no difference between the weak interaction and electromagnetism—the two forces are symmetrical, in the gauge sense. The combined interaction 'works' under conditions like those early on in the Big Bang. But as the Universe cooled, the symmetry got broken, and the two forces went their separate ways.

How can a symmetry get broken? Think of a bar of magnetic material, containing an enormous number of individual tiny magnets, each the equivalent of one of the atoms in the bar. The behaviour of those tiny magnets depends on the temperature of the bar. Heat is a form of energy, and the atoms—or tiny magnets—in a hot bar have more energy than those in a cold bar. Their energy is in the form of motion. The hotter the bar, the more its atoms jostle against one another, and if the bar gets very hot they jostle about so much that the bar melts and the component atoms move more freely about in the liquid form. But we don't have to heat a bar of magnetic material that much to learn about broken symmetry.

When the magnetic material is hot, the tiny internal magnets spin around at random and point in all directions. There is, then, no overall magnetism associated with the bar. But when the bar is cooled below a certain temperature, called the Curie temperature, it suddenly takes up a magnetized state in which all the internal magnets line up with one another and point the same way. The bar has a definite magnetization. At high temperature, the lowest 'energy state' of the bar is with all the magnets pointing at random. At lower temperatures, the lowest energy state is with the magnets aligned, but it doesn't matter which way round they line up. The high-temperature state is symmetric, but the low-temperature state is asymmetric. The cool bar has a preferred orientation, and the symmetry is broken. The change occurred because, at high temperatures, the thermal energy of the atoms overcame the magnetic forces. At low temperatures, on the other hand, magnetic forces overwhelmed the thermal agitation of the atoms.

According to the theory of inflation, proposed by Alan Guth of the Massachusetts Institute of Technology something similar happened early in the life of the Universe. When the Universe was very hot and very dense, in the first

10^{-35} of a second, electromagnetism, the weak interaction and the strong interaction were all united as one superforce, a symmetric interaction. But as the Universe began to cool, the symmetry was broken, and the three forces went their separate ways. Clearly, the two states of the Universe—before and after symmetry breaking—are different from each other. The change from one state to the other turns out to be similar in some ways to the change from steam to liquid water or from water to ice: it is a phase transition. And, like those phase transitions when water cools, the symmetry-breaking phase transition of the early Universe must have released energy. According to the inflationary scenario, this sudden release of energy acted like negative gravity, inflating the Universe and blasting everything apart in a fraction of a second. This brief phase of rapid expansion occurred exponentially, the theorists believe, doubling the size of the Universe every 10^{-35} of a second or so. Such a rapid inflation would have taken the entire Universe as we know it from the size of a proton to the size of a basketball in a fraction of a second, and all well before the end of the first hundredth of a second of the life of the Universe. After that, a few billion years of more modest, steady expansion would have sufficed to make the observable Universe as big as we see it today.

You think you know how small an atom is? A proton is a lot smaller than that. It is about one hundred thousandth (10^{-5}) of the size of an atom. A hundred thousand protons laid side by side would stretch across the diameter of an atom. In round terms, it would take the *cube* of this number, 10^{15} protons, to fill up the space occupied by a single atom. That's one proton for every couple of minutes of the time the Universe as we know it has been in existence—and that would be to 'fill up' just *one* atomic volume. We already know how small an atomic volume itself is. Inflating a basketball to the size of the observable Universe is easy if you have 15 billion years to do it. It is inflating something the size of a proton to the size of a basketball that is tricky. But the best modern theory of particle physics and the Universe tells us that that is exactly what happened, in a split second, just about 10^{-35} of a second after the moment of creation.

Particle physics and cosmology make strange bedfellows. But together they provide a completely self-consistent and logical description of how the Universe got to be the way we see it today, a description which is in line with all of the laws of physics determined by studying collisions between particles smaller than atoms in accelerators like the ones at CERN. You can't ask much more of any scientific theory.

seventeen

Inflation for Beginners

▼

INFLATION HAS BECOME A cosmological buzzword in the 1990s. No self-respecting theory of the Universe is complete without a reference to inflation—and at the same time there is now a bewildering variety of different versions of inflation from which to choose. Clearly, what's needed is a beginner's guide to inflation, with which newcomers to cosmology can find out just what this exciting development is all about. This is it—new readers start here.

The reason why something like inflation was needed in cosmology was highlighted by discussions of two key problems in the 1970s. The first of these is the horizon problem—the puzzle that the Universe looks the same on opposite sides of the sky (opposite horizons), even though there has not been time since the Big Bang for light (or anything else) to travel across the Universe and back. So how do the opposite horizons 'know' how to keep in step with each other? The second puzzle is called the flatness problem: the spacetime of the Universe is very nearly flat, which means that the Universe sits just on the dividing line between eternal expansion and eventual recollapse.

The flatness problem can be understood in terms of the density of the Universe. The density parameter is a measure of the amount of gravitating material in the Universe, usually denoted by the Greek letter omega (Ω), and also known as the flatness parameter. It is defined in such a way that if spacetime is exactly flat, then $\Omega = 1$. Before the development of the idea of inflation, one of the great puzzles in cosmology was the fact that the actual density of the Universe today is very close to this critical value—certainly within a factor of 10. This is curious because, as the Universe expands away from the Big Bang, the expansion will push the density parameter away from the critical value. If the Universe starts out with the parameter less than 1, $\Omega =$ gets smaller as the Universe ages, while if it starts out bigger than 1, Ω gets bigger as the Universe ages. The fact that Ω is between 0.1 and 1 today means that in the first second of the Big Bang it was precisely 1 to within 1 part in 10^{60}. This makes the value of the density parameter in the beginning one of the most precisely determined numbers in all of science, and the natural inference is that the value is, and always has been, exactly 1. One important implication of this is that there must be a large amount of dark matter in the Universe. Another is that the Universe was made flat by inflation.

Inflation is a general term for models of the very early Universe which involve a short period of extremely rapid (exponential) expansion, blowing the size of what is now the observable Universe up from a region far smaller than a proton to about the size of a basketball in a small fraction of a second. This process would smooth out spacetime to make the Universe flat, and would also resolve the horizon problem by taking regions of space that were once close enough to have got to know each other well and spreading them far apart, on opposite sides of the visible Universe today.

Inflation became established as the standard model of the very early Universe in the 1980s. It achieved this success not only because it resolves many puzzles about the nature of the Universe, but because it did so using the grand unified theories (GUTs) and understanding of quantum theory developed by particle physicists completely independently of any cosmological studies. These theories of the particle world had been developed with no thought that they might be applied in cosmology (they were in no sense 'designed' to tackle all the problems they turned out to solve), and their success in this area suggested to many people that they must be telling us something of fundamental importance about the Universe.

The marriage of particle physics (the study of the very small) and cosmology (the study of the very large) seems to provide an explanation of how the Universe began, and how it got to be the way it is. Inflation is therefore regarded as the most important development in cosmological thinking since the discovery that the Universe is expanding first suggested that it began in a Big Bang.

Taken at face value, the observed expansion of the Universe implies that it was born out of a singularity, a point of infinite density, some 15 billion years ago. (Cosmologists still disagree about exactly how old the Universe is, but the exact age doesn't affect the argument.) Quantum physics tells us that it is meaningless to talk in quite such extreme terms, and that instead we should consider the expansion as having started from a region no bigger across than the so-called Planck length (10^{-35} m), when the density was not infinite but 'only' some 10^{94} grams per cubic centimetre. These are the absolute limits on size and density allowed by quantum physics.

On that picture, the first puzzle is how anything that

dense could ever expand—it would have an enormously strong gravitational field, turning it into a black hole and snuffing it out of existence (back into the singularity) as soon as it was born. But it turns out that inflation can prevent this from happening, while quantum physics allows the entire Universe to appear, in this supercompact form, out of nothing at all, as a cosmic free lunch. The idea that the Universe may have appeared out of nothing at all, and contains zero energy overall, was developed by Edward Tryon, of the City University in New York, who suggested in the 1970s that it might have appeared out of nothing as a so-called vacuum fluctuation, allowed by quantum theory.

Quantum uncertainty allows the temporary creation of bubbles of energy, or pairs of particles (such as electron—positron pairs) out of nothing, provided they disappear in a short time. The less energy is involved, the longer the bubble can exist. Curiously, the energy in a gravitational field is negative, while the energy locked up in matter is positive. If the Universe is exactly flat, then, as Tryon pointed out, the two numbers cancel out, and the overall energy of the Universe is precisely zero. In that case, the quantum rules allow it to last forever.

If you find this mind-blowing, you are in good company. George Gamow told in his book *My World Line* how he was having a conversation with Albert Einstein while walking through Princeton in the 1940s. Gamow casually mentioned that one of his colleagues had pointed out to him that, according to Einstein's equations, a star could be created out of nothing at all, because its negative gravitational energy precisely cancels out its positive mass energy. 'Einstein stopped in his tracks,' says Gamow, 'and, since we were crossing a street, several cars had to stop to avoid running us down.'

Unfortunately, if a quantum bubble (about as big as the Planck length) containing all the mass-energy of the universe (or even a star) did appear out of nothing at all, its intense gravitational field would (unless something else intervened) snuff it out of existence immediately, crushing it into a singularity. So the free-lunch Universe seemed at first like an irrelevant speculation. However, as with the problems involving the extreme flatness of spacetime, and its appearance of extreme homogeneity (the same everywhere) and isotropy (the same in all directions, most clearly indicated by the uniformity of the background radiation), the development of the inflationary scenario showed how to remove this difficulty and allow such a quantum fluctuation to expand exponentially up to macroscopic size before gravity could crush it out of existence.

All these problems would be resolved if something gave the Universe a violent outward push (in effect, acting like antigravity) when it was still about a Planck length in size. Such a small region of space would be too tiny, initially, to contain irregularities, so it would start off homogeneous and isotropic. There would have been plenty of time for signals travelling at the speed of light to have criss-crossed the ridiculously tiny volume, and so there is no horizon problem—both sides of the embryonic universe are 'aware' of each other. And spacetime itself gets flattened by the expansion, in much the same way that the wrinkly surface of a prune becomes a smooth, flat surface when the prune is placed in water and swells up. As in the standard Big Bang model, we can still think of the Universe as like the skin of an expanding balloon, but now we have to think of it as an absolutely enormous balloon that was hugely inflated during the first split second of its existence.

The reason why the GUTs created such a sensation when they were applied to cosmology is that they predict the existence of exactly the right kind of mechanisms to do this trick. They are called scalar fields, and they are associated with the splitting apart of the original grand unified force into the fundamental forces we know today, as the Universe began to expand and cool. Gravity itself would have split off at the Planck time, 10^{-43} of a second, and the strong nuclear force by about 10^{-35} of a second. Within about 10^{-32} of a second, the scalar fields would have done their work, doubling the size of the Universe at least once every 10^{-34} of a second (some versions of inflation suggest even more rapid expansion than this).

This may sound modest, but it would mean that in 10^{-32} of a second there were 100 doublings. This rapid expansion is enough to take a quantum fluctuation 10^{20} times smaller than a proton and inflate it to a sphere about 10 cm across in about 15×10^{-33} of a second. At that point, the scalar field has done its work of kick-starting the Universe and is settling down, giving up its energy and leaving a hot fireball expanding so rapidly that, even though gravity can now begin to do its work of pulling everything back into a Big Crunch, it will take hundreds of billions of years to first halt the expansion and then reverse it.

Curiously, this kind of exponential expansion of space-time is exactly described by one of the first cosmological models developed using the General Theory of Relativity, by Willem de Sitter in 1917. For more than half a century, this de Sitter model seemed to be only a mathematical curiosity, of no relevance to the real Universe, but it is now one of the cornerstones of inflationary cosmology.

When the General Theory of Relativity was published in 1916, de Sitter reviewed the theory and developed his own ideas in a series of three papers which he sent to the Royal Astronomical Society in London. The third of these papers included discussion of possible cosmological models—both what turned out to be an expanding Universe (the first model of this kind to be developed, although the implications were not fully appreciated in 1917) and an oscillating Universe model.

De Sitter's solution to Einstein's equations seemed to describe an empty, static Universe (empty spacetime). But in the early 1920s it was realized that if a tiny amount of matter was added to the model (in the form of particles scattered throughout the spacetime), they would recede from each other exponentially fast as the spacetime expanded. This means that the distance between two particles would double repeatedly on the same timescale, so they would be twice as far apart after one tick of some cosmic clock, four times as far apart after two ticks, eight times as far apart after three ticks, sixteen times as far apart after four ticks, and so on. It would be as if each step you took down the road took you twice as far as the previous step.

This seemed to be completely unrealistic; even when the expansion of the Universe was discovered, later in the 1920s, it turned out to be much more sedate. In the expanding Universe as we see it now, the distances between 'particles' (clusters of galaxies) increase steadily—they take one step for each tick of the cosmic clock. The distance is increased by a total of two steps after two ticks, three steps after three ticks, and so on. In the 1980s, however, when the theory of inflation suggested that the Universe really did experience a stage of exponential expansion during the first split second after its birth, this inflationary exponential expansion turned out to be exactly described by the de Sitter model, the first successful cosmological solution to Einstein's equations of the General Theory of Relativity.

One of the peculiarities of inflation is that it seems to take place faster than the speed of light. Even light takes 30 billionths of a second (3×10^{-10} of a second) to cross a single centimetre, and yet inflation expands the Universe from a size much smaller than a proton to 10 cm across in only 15×10^{-33} of a second. This is possible because it is spacetime itself that is expanding, carrying matter along for the ride. Nothing is moving through spacetime faster than light, either during inflation or ever since. Indeed, it is just because the expansion takes place so quickly that matter has no time to move while it is going on, and the process 'freezes in' the original uniformity of the primordial quantum bubble that became our Universe.

The inflationary scenario has already gone through several stages of development during its short history. The first inflationary model was developed by Alexei Starobinsky, at the L. D. Landau Institute of Theoretical Physics in Moscow, at the end of the 1970s—but it was not then called 'inflation'. It was a very complicated model based on a quantum theory of gravity, but it caused a sensation among cosmologists in what was then the Soviet Union, becoming known as the 'Starobinsky model' of the Universe. Unfortunately, because of the difficulties Soviet scientists still had in travelling abroad or communicating with colleagues outside the Soviet sphere of influence at that time, the news did not spread outside their country.

In 1981, Alan Guth, then at the Massachusetts Institute of Technology, published a different version of the inflationary scenario, not knowing anything of Starobinsky's work. This version was more accessible in both senses of the word—it was easier to understand, and Guth was based in the USA, able to discuss his ideas freely with colleagues around the world. And as a bonus, Guth came up with the catchy name 'inflation' for the process he was describing. There were obvious flaws with the specific details of Guth's original model (which he acknowledged at the time). In particular, Guth's model left the Universe after inflation filled with a mess of bubbles, all expanding in their own way and colliding with one another. We see no evidence for these bubbles in the real Universe. Obviously, the simplest model of inflation couldn't be right, but it was this version of the idea that made every cosmologist aware of the power of inflation.

In October 1981, there was an international meeting in Moscow at which inflation was a major talking point. Stephen Hawking presented a paper claiming that inflation could not be made to work at all, but the Russian cosmologist Andrei Linde presented an improved version, called 'new inflation', which got round the difficulties with Guth's model. Ironically, Linde was the official translator for Hawking's talk, and had the embarrassing task of offering the audience the counter-argument to his own work! But after the formal presentations Hawking was persuaded that Linde was right, and inflation might be made to work after all. Within a few months, the new inflationary scenario was also published by Andreas Albrecht and Paul Steinhardt of the University of Pennsylvania, and by the end of 1982 inflation was well established.

Linde has been involved in most of the significant developments with the theory since then. The next step forward came with the realization that there need not be anything special about the Planck-sized region of spacetime that expanded to become our Universe. If that was part of some larger region of spacetime in which all kinds of scalar fields were at work, then only the regions in which those fields produced inflation could lead to the emergence of a large Universe like our own. Linde called this 'chaotic inflation', because the scalar fields can have any value at different places in the early super-universe. It is the standard version of inflation today (but note that this use of the term 'chaos' is like the everyday meaning implying a complicated mess, and has nothing to do with the mathematical subject known as 'chaos theory').

The idea of chaotic inflation led to what is (so far) the ultimate development of the inflationary scenario. The great unanswered question in standard Big Bang cosmology is what came 'before' the singularity. It is often said that the question is meaningless, since time itself began at the singularity. But chaotic inflation suggests that our Universe grew out of a quantum fluctuation in some pre-existing region of spacetime, and that exactly equivalent processes can create regions of inflation within our own Universe. In effect, new universes bud off from our Universe, and our Universe may itself have budded off from another universe, in a process which had no beginning and will have no end. A variation on this theme suggests that the 'budding' process takes place through black holes, and that every time a black hole collapses into a singularity it 'bounces' out into another set of spacetime dimensions, creating a new inflationary universe. This is called the baby universe scenario.

There are similarities between the idea of eternal inflation and a self-reproducing universe and the version of the steady state hypothesis developed in England by Fred Hoyle and Jayant Narlikar, with their C-field playing the part of the scalar field that drives inflation. As Hoyle wryly pointed out at a meeting of the Royal Astronomical Society in London in December 1994, the relevant equations in inflation theory are exactly the same as in his version of the steady state idea, but with the letter 'C' replaced by the Greek 'Φ'. 'This,' said Hoyle (tongue firmly in cheek), 'makes all the difference in the world.'

Modern proponents of the inflationary scenario arrived at these equations entirely independently of Hoyle's approach, and are reluctant to accept this analogy, having cut their cosmological teeth on the Big Bang model. Indeed, when Guth was asked in 1980 how the then-new idea of inflation related to the steady state theory, he is reported as replying, 'What is the steady state theory?' But although inflation is generally regarded as a development of Big Bang cosmology, it is better seen as marrying the best features of both the Big Bang and the steady state scenario.

This might all seem like a philosophical debate as futile as the argument about how many angels can dance on the head of a pin, except for the fact that observations of the background radiation by COBE showed exactly the pattern of tiny irregularities that the inflationary scenario predicts. One of the first worries about the idea of inflation (long ago in 1981) was that it might be too good to be true. In particular, if the process was so efficient at smoothing out the Universe, how could irregularities as large as galaxies, clusters of galaxies and so on ever have arisen? But when the researchers looked more closely at the equations they realized that quantum fluctuations should still have been producing tiny ripples in the structure of the Universe even when our Universe was only something like 10^{-25} of a centimetre across—a hundred million times bigger than the Planck length.

The theory said that inflation should have left behind an expanded version of these fluctuations, in the form of irregularities in the distribution of matter and energy in the Universe. These density perturbations would have left an imprint on the background radiation at the time matter and radiation decoupled (about 300,000 years after the Big Bang), producing exactly the kind of non-uniformity in the background radiation that has now been seen, initially by COBE and later by other instruments. After decoupling, the density fluctuations grew to become the large-scale structure of the Universe revealed today by the distribution of galaxies. This means that the COBE observations are actually giving us information about what was happening in the Universe when it was less than 10^{-20} of a second old.

No other theory can explain both why the Universe is so uniform overall and yet contains exactly the kind of 'ripples' represented by the distribution of galaxies through space and by the variations in the background radiation. This does not prove that the inflationary scenario is correct, but it is worth remembering that had COBE found a different pattern of fluctuations (or no fluctuations at all), that would have proved the inflationary scenario wrong. In the best scientific tradition, the theory made a major and unambiguous prediction which did 'come true'. Inflation also predicts that the primordial perturbations may have left a trace in the form of gravitational radiation with particular characteristics, and it is hoped that detectors sensitive enough to identify this characteristic radiation may be developed within the next ten or twenty years.

The clean simplicity of this simple picture of inflation has now, however, begun to be obscured by refinements, as inflationary cosmologists add bells and whistles to their models to make them match more closely the Universe we see about us. Some of the bells and whistles, it has to be said, are studied just for fun. Linde himself has taken great delight in pushing inflation to extremes, and offering entertaining new insights into how the Universe might be constructed. For example, could our Universe exist on the inside of a single magnetic monopole produced by cosmic inflation? According to Linde, it is at least possible, and may be likely. And in a delicious touch of irony, Linde, who now works at Stanford University, made this outrageous claim in a lecture at a workshop on the birth of the Universe held recently in Rome, where the view of 'creation' is usually rather different.

One of the reasons why theorists came up with the idea of inflation in the first place was precisely to get rid of magnetic monopoles—strange particles carrying isolated north or south magnetic fields, predicted by many Grand Unified Theories of physics but never found in nature. Standard models of inflation solve the 'monopole problem' by arguing that the seed from which our entire visible Universe grew was a quantum fluctuation so small that it only contained one monopole. That monopole is still out there, somewhere in the Universe, but it is highly unlikely that it will ever pass our way.

But Linde has discovered that, according to theory, the conditions that create inflation persist *inside* a magnetic monopole, even after inflation has halted in the Universe at large. Such a monopole would look like a magnetically charged black hole, connecting our Universe through a wormhole in spacetime to another region of inflating spacetime. Within this region of inflation, quantum processes can produce monopole—antimonopole pairs, which then separate exponentially rapidly as a result of the inflation. Inflation then stops, leaving an expanding Universe rather like our own which may contain one or two monopoles, within each of which there are more regions of inflating spacetime.

The result is a never-ending fractal structure, with inflating universes embedded inside each other and connected through the magnetic monopole wormholes. Our Universe may be inside a monopole which is inside another universe which is inside another monopole, and so on indefinitely. What Linde calls 'the continuous creation of exponentially expanding space' means that 'monopoles by themselves can solve the monopole problem'. Although it seems bizarre, the idea is, he stresses, 'so simple that it certainly deserves further investigation'.

That variation on the theme really is just for fun, and it is hard to see how it could ever be compared with observations of the real Universe. But most of the modifications to inflation now being made are in response to new observations, and in particular to the suggestion that spacetime may not be quite 'flat' after all. In the mid-1990s, many studies (including observations made by the repaired Hubble Space Telescope) began to suggest that there might not be quite enough matter in the Universe to make it perfectly flat—most of the observations suggest that there is only 20 or 30 per cent as much matter around as the simplest versions of inflation require. The shortfall is embarrassing, because one of the most widely publicized predictions of simple inflation was the firm requirement of exactly 100 per cent of this critical density of matter. But there are ways around the difficulty. Here are two.

The first suggestion is almost heretical, in the light of the way astronomy has developed since the time of Copernicus. Is it possible that we are living near the centre of the Universe? For centuries, the history of astronomy has seen human beings displaced from any special position. First, the Earth was seen to revolve around the Sun; then the Sun was seen to be an insignificant member of the Milky Way Galaxy; then the Galaxy was seen to be an ordinary member of the cosmos. But now comes the suggestion that the 'ordinary' place to find observers like us may be in the middle of a bubble in a much greater volume of expanding space.

The conventional version of inflation says that our entire visible Universe is just one of many bubbles of inflation, each doing its own thing somewhere out in an eternal sea of chaotic inflation. The process of rapid expansion, however, forces spacetime in all the bubbles to be flat. The bubbles that form in a bottle of fizzy cola when the top is opened is a useful analogy. But that suggestion, along with other cherished cosmological beliefs, has now been challenged by Linde, working with his son Dmitri Linde (of Caltech) and Arthur Mezhlumian (also of Stanford).

Linde and his colleagues point out that the Universe in which we live is like a hole in a sea of superdense, exponentially expanding inflationary cosmic material, within which there are other holes. All kinds of bubble universes will exist, and it is possible to work out the statistical nature of their properties. In particular, the two Lindes and Mezhlumian have calculated the probability of finding yourself in a region of this super-universe with a particular density—for example, the density of 'our' Universe.

Because very dense regions blow up exponentially quickly (doubling in size every fraction of a second), it turns out that the volume of all regions of the super-universe with twice any chosen density is 10 to the power of 10 million times greater than the volume of the super-universe with the chosen density. For any chosen density, most of the matter at that density is near the middle of an expanding bubble, with a concentration of more dense material round the edge of the bubble. But even though some of the higher-density material is round the edges of low-density bubbles, there is even more (vastly more!) higher-density material in the middle of higher-density bubbles, and so on for ever.

The discovery of this variation on the theme of fractal structure surprised the researchers so much that they confirmed it by four independent methods before venturing to announce it to their colleagues. Because the density distribution is non-uniform on the appropriate distance scales, it means that not only may we be living near the middle of a bubble Universe, but that the density of the region of space we can see may be less than the critical density, compensated for by extra density beyond our field of view.

This is convenient, since those observations by the Hubble Space Telescope have suggested that cosmological models which require exactly the critical density of matter may be in trouble. But there is more. Those Hubble observations assume that the parameter which measures the rate at which the Universe is expanding, the Hubble constant, really is a constant, the same everywhere in the observable Universe. If Linde's team is right, however, the measured value of the 'constant' may be different for galaxies at different distances from us, truly throwing the cat among the cosmological pigeons. We may seem to live in a low-density Universe in which both the measured density and the value of the Hubble constant will depend on which volume of the Universe these properties are measured over!

That would mean abandoning many cherished ideas about the Universe, and may be too much for many cosmologists to swallow. But there is a simpler solution to the density puzzle, one which involves tinkering only with the models of inflation, not with long-held and cherished cosmological beliefs. That may make it more acceptable to most cosmologists—and it's so simple that it falls into the 'why didn't I think of that?' category of great ideas.

A double dose of inflation may be something to make the Government's hair turn grey—but it could be just what cosmologists need to rescue their favourite theory of the origin of the Universe. By turning inflation on twice, they have found a way to have all the benefits of the inflationary scenario, while still leaving the Universe in an 'open' state, so that it will expand forever.

In those simplest inflation models, remember, the big snag is that after inflation even the observable Universe is left like a mass of bubbles, each expanding in its own way. We see no sign of this structure, which has led to all the refinements of the basic model. Now, however, Martin Bucher and Neil Turok of Princeton University, working with Alfred Goldhaber of the State University of New York, have turned this difficulty to advantage.

They suggest that after the Universe had been homogenized by the original bout of inflation, a second burst of inflation could have occurred within one of the bubbles. As inflation begins (essentially at a point), the density is effectively 'reset' to zero, and rises towards the critical density as inflation proceeds and energy from the inflation process is turned into mass. But because the Universe has already been homogenized, there is no need to require this bout of inflation to last until the density reaches the critical value. It can stop a little sooner, leaving an open bubble (what we see as our entire visible Universe) to carry on expanding at a more sedate rate. They actually use what looked like the fatal flaw in Guth's model as the basis for their scenario.

According to Bucher and his colleagues, an end product looking very much like the Universe in which we live can arise naturally in this way, with no 'fine tuning' of the inflationary parameters. All they have done is use the very simplest possible version of inflation, going back to Alan Guth's work, but they've applied it twice. And you don't have to stop there. Once any portion of expanding spacetime has been smoothed out by inflation, new inflationary bubbles arising inside that volume of spacetime will all be pre-smoothed and can end up with any amount of matter from zero to the critical density (but no more). This should be enough to make everybody happy. Indeed, the biggest problem now is that the vocabulary of cosmology doesn't quite seem adequate to the task of describing all this activity.

The term 'Universe' is usually used for everything of which we can ever have knowledge: the entire span of space and time accessible to our instruments, now and in the future. This may seem like a fairly comprehensive definition, and in the past it has traditionally been regarded as synonymous with the entirety of everything that exists. But the development of ideas such as inflation suggests that there may be something else beyond the boundaries of the observable Universe—regions of space and time that are unobservable in principle, not just because light from them has not yet had time to reach us or because our telescopes are not sensitive enough to detect their light.

This has led to some ambiguity in the use of the term 'Universe'. Some people restrict it to the observable Universe, while others argue that it should be used to refer to all of space and time. If we use 'Universe' as the name for our own expanding bubble of spacetime, everything that is in principle visible to our telescopes, then maybe the term 'cosmos' can be used to refer to the entirety of space and time, within which (if the inflationary scenario is correct) there may be an indefinitely large number of other expanding bubbles of spacetime, other universes with which we can never communicate. Cosmologists ought to be happy with the suggestion, since it makes their subject infinitely bigger and therefore infinitely more important.

Further Reading

▼

If you want to know which of my own books relate to the stories told here, the following checklist will point you in the right direction:

Chapter One and Chapter Two
Fire on Earth (with Mary Gribbin; Simon & Schuster, 1996)
Chapter Three
Being Human (with Mary Gribbin; Phoenix, 1993)
Chapters Four to Seven
Blinded by the Light (Harmony, 1991)
Chapter Seven and Chapter Ten
Companion to the Cosmos (Phoenix, 1996)
Chapters Eight, Ten and Seventeen
In the Beginning (Penguin, 1993)
The Birth of Time (Weidenfeld & Nicolson, 1999)
Chapter Nine and Chapter Twelve
In Search of the Edge of Time (Penguin, 1992)
Chapter Eleven and Chapter Sixteen
In Search of the Big Bang (revised edition, Penguin, 1998)

Chapter Thirteen
White Holes (Paladin, 1977)
Chapter Fourteen
The Omega Point (Bantam, 1987)
Chapter Fifteen
The Matter Myth (with Paul Davies, Penguin, 1991)

If you want other guides to the excitement of Universe-watching, the following are recommended:

Simon Goodwin, *Hubble's Universe* (Constable, 1996)
Robert Jastrow, *God and the Astronomers* (Norton, 1994)
Martin Rees, *Before the Beginning* (Simon & Schuster, 1997)
Joseph Silk, *A Short History of the Universe* (*Scientific American*/W. H. Freeman, 1994)
Steven Weinberg, *The First Three Minutes* (Deutsch, 1977)

Index

▼